Ship Management Ser

Managing Ships

By JOHN M. DOWNARD

Fairplay Publications

Published and distributed by
FAIRPLAY PUBLICATIONS LTD
52/54 Southwark Street, London SE1 1UJ
Telephone 403 3395
Telex 884595 FPLAY G

ISBN 0 905 045 59 9

Printed by Page Bros (Norwich) Ltd,
Mile Cross Lane, Norwich, Norfolk NR6 6SA

The Author

The author was born in 1928 and began his career with the P & O Group in 1944. His sea service was spent in tramp ships and bulk carriers and included a period as a hull inspector and eight years in command. He came ashore in 1968 and held positions as Assistant Marine Superintendent, Fleet Personnel Manager and Assistant Fleet Manager.

He joined the New York based Fairfield Maxwell Group in 1975 as Managing Director of their London ship management company. From 1979 to 1981 he was a Director of a shipbroking and liner agency company and a marine consultant within the group. He was also a Director of the UK subsidiary of Marine Management Systems Inc. In 1981 he was appointed an Area Director for Reefer Express Lines Pty Ltd.

Author's Preface

I am most grateful to Brian Hill, Chris Hewer and William Packard who so kindly gave me their constructive comments on various chapters of this book.

I am also very grateful to:

The Egyptian Maritime Transport Academy, Alexandria for allowing me to use their reference library.

M. H. Smith for his permission and advice in regard to organisational diagrams.

The P & O Steam Navigation company for their permission to use extracts from the Hain-Nourse Limited Sea Staff Positions Manual.

The International Shipping Federation Ltd for their permission to reproduce the text of the ICS/ISF Code of Good Management Practice in Safe Ship Operation.

M.R. Holman for his permission to use quotations from the Handy Book for Shipowners and Masters.

And finally to Corri Roos and Josianne André for their patient efforts in typing drafts and the final manuscripts.

The front cover depicts the Brazilian built 26,500 dwt bulker *Cape Finisterre*, photographed in the Dover Strait by FotoFlite.

Dedication

To my Mother and Father

Table of Contents

PART TWO

PART THREE

Introduction

There are many excellent books on the subject of "Management" and many of the philosophies, theories and facts contained in them are applicable to "Ship Management". For this reason points of theory, etc., are kept to a minimum in this book except as they apply directly to "Ship Management".

This book is intended as a companion to "Running Costs" and although, inevitably, there is some overlap between the subjects of Managing Ships and the Costs of Running them, this is also kept to a minimum.

As with "Running Costs" it is necessary to clarify some meanings of words at the beginning as follows:

Ship management: The functions of taking care of a ship, i.e. responsibility for manning, maintaining, supplying and insuring the ship, and ensuring that the ship is available to the operators for the maximim amount of time possible. In other words, all the activities not carried out by the operators.

Operations: The functions associated with the earnings of a ship, i.e. responsibility for obtaining cargoes, scheduling, stemming or ordering bunkers, making arrangements for the loading and discharge of cargoes and associated port activities and the lay up of ships. The term also includes the sale and purchase and chartering of ships.

Running Costs: The costs of managing the ship, i.e. the costs of all the activities associated with management of the ship.

Accountable: To be accountable in the management sense is to be responsible for actions and **results**. To carry out such responsibility a manager must have the necessary authority.

Responsible: To be responsible is to be answerable for an assignment for which authority has been given.

Policy: Is a selected line of conduct formally adapted by a Company such that individual decisions can be made and a common purpose achieved.

Line and Staff: These are management terms originally associated with army organisation. The line managers have responsibility for the total product which in the case of ship management is the ship. Staff managers provide the necessary support for the ship and in this book are referred to as functional managers.

Centralisation: Is a system of management which concentrates decisions at the top or center of the organisation. The divisions or departments concentrate on their functions rather than on the business as a whole.

Federal Decentralisation: Is a system of operating a group of companies or parts of a company like a federation, i.e. a central government or top management with a number of self governing autonomous businesses to whom full accountability is delegated.

Its benefits are that it allows everyone in the organisation to concentrate on performance and results, including the top management whose tasks are different to those managing the business.

Simulated Decentralisation: Not all units, groups, etc., can be set up as a business in their own right, although they may be much more complex than a functional department. In such cases it is possible to simulate decentralisation by producing "costs" or "earnings' although these may be outside the direct influence of the unit or group.

An example of this is the charters or freight earnings of ships. These are obtained by the operators but kept in mind by the ship managers in their endeavours to keep costs contained such that profits will result.

Since the 1950s, concepts of ship management have undergone considerable change as have the tasks of the people engaged in managing them. In many cases traditions have had to be cast aside and change forced upon everyone involved in ships because of economic pressures.

In some countries those same economic pressures, which essentially are the inability to make a profit because costs are too high and earnings too low, or both, have resulted in a considerable reduction in some national fleets. Those who have been able to stay in business have been searching continuously for ways of managing ships more efficently and thus reduce costs.

There are now a number of ways in which ships can be managed and shipowners do have some choice in the way they wish to care for their investment. This book is about those different ways and the elements of ship management which are common to them all.

Chapter One

Management and shipping

"Business is a human organisation made or broken by its people"

Peter Drucker

Management

To the newcomer there are a bewildering number of books on management. All to a greater or lesser degree describe the tools of management; planning, organising, controlling, communicating, the disciplines and the restraints put upon it.

Some describe the styles of management; autocratic, democratic, bureaucratic, entrepreneural, creative etc. Others write, semi-humourously, of types of management such as *Management by Crisis*, by default, by procrastination, by harrassment and by Machiavelli. Yet again others write of techniques and philosophies of management such as *Management by Objectives*. Peter Drucker, one of the most famous management authors writes of *Managing for Results* and *Effective Management* and even stresses the need for managers to first learn how to manage themselves before managing others.

Management has acquired many definitions through the years but a common factor in all is "responsibility", regardless of whether or not other people are involved. In most cases management is about being responsible for something with and through other people.

Although not everyone is suited to be a manager, the old idea of people being born managers has been put aside and today many are effectively trained and developed into the job. As most management involves working with others, the ability to lead and to be a good team man is essential, but the most important attribute is that of "judgement", the ability to make the right decision at the right time. Today this sense of judgement, usually gained through experience and training, is demanded beyond the normal business requirements as sophisticated aids to business are rapidly developed. In this the manager almost requires a crystal ball to see his future requirements as expensive systems installed today are outdated tomorrow.

Many managers find themselves spending much of their time in personnel matters and, in rapidly changing situations, the problem for themselves and their staff is coping with "change". They also find themselves involved with the "Institutions"; Government Departments, Employers and Industrial Federations, Professional Associations, Educational systems, Unions, and in some cases International Organisations. Sometimes the manager has a choice as to the amount of his involvement with them and his business, but at other times he can only be cooperative with them to a greater or lesser degree depending on his opinion as to their value and achievable results.

Management in shipping

In the last days of sail and early days of steam, ships were, in the main, self managed. Communications were so poor that the shipowner had no option but to trust his shipmasters once the ships were out of sight of the home port. Of necessity the master had to make all the short term decisions including those associated with the employment of the ship. As communications improved it became easier to instruct and thus control ships; and much of the decision making on major technical, supplies, and crew matters, moved from the ships to the head office. Thus it can be said that the system of management changed from a primitive decentralised system to a centralised system.

This shift in the decision making process occurred with the development of the functional departments and became the standard style of ship management which existed until the 1960s. The traditional methods of working in the shipping industry at that time were very strong, as they were in many other industries, and resistance to change was equally strong. However, as so often happens with organisations and people, change was forced upon the older shipping companies by events rather than choice.

In the older British, North European and Scandinavian shipping companies the change was brought about to a large degree by the increased costs of running ships, low freights, and increased competition from the new maritime nations. There were also difficulties in keeping men in ships. It seemed that the enthusiasm for the sea still existed but men were soon dissatisfied with the life and left, not considering it to be a worthwhile long term career.

The first instinctive action was to try and cut costs as had been done so many times before, particularly in the tramp trades at the bottom of the trading cycles. This resulted in even more central control and from the point of view of the sea staff, some of the less attractive techniques of management started to creep in; more reports, more demands for quantitive information about every aspect of the ship's performance and inevitably, more and more instructions, regulations, etc. There was also re-thinking about the frequency of drydocking, hull painting, the amounts of stock carried and consideration as to how and where purchases were made.

There is no doubt that these actions had some effect, but it was not enough and attention turned towards crew costs which had become the largest single item, although there had been an increase in all other costs including fuel and lubricants.

At first it was thought that reducing the number of crew in ships was the answer, but this was a very emotive subject. Technical staff ashore and on board protested that maintenance would fall behind and thus they needed to be convinced that reductions could be made. The unions also protested about the loss of employment opportunities for their members. But it was not just the numbers of crew employed which was causing the crew costs to be high. The high turnover of staff was also costly because of the wasted training and development of staff, the necessary recruitment of replacements and the resultant increase in the Personnel Department ashore to cope with the problems.

Faced with this situation a number of leading shipowners in Britain, North Europe and Scandinavia, independently and in cooperation with a number of industrial and professional societies, institutes and federations, sought answers to the problems. At first this was done in two related ways: By arranging studies of the work done in ships and by developing the sea and shore staff by sending them on courses and seminars.

The studies of ships showed that crews could be reduced by work being done more effectively, either through a change in methods and/or the re-design of the working areas and equipment. Some maintenance was found to be excessive while some was inadequate and it was often found that there was wastage of equipment and supplies which could be corrected. These studies resulted in the development of new systems of work in ships, such as General Purpose and Interdepartmental Flexibility Crews, Planned Maintenance, Spare gear control and Stock control systems. They also resulted in an improved mix of the skills in the crew to suit optimum technical and operational requirements. Significantly the new systems showed the need for the sea staff to cooperate for the good of the ship as a whole and not just their own departments, that is, there was a requirement for a team management.

They also showed that there were a number of social problems in ships which were, essentially, that there was not enough "job satisfaction" in going to sea. The feedback from the various courses, seminars, etc. gave a constant cry for more information, more involvement and more responsibility. There was also a frequently expressed concern about alcoholism which may have resulted from people with too much time on their hands and not enough stimulus.

But the studies also drew attention to one very important fact: That the ship was part of the total company organisation and thus it was wrong to study the ship organisation in isolation. If a study was to be made effectively the whole organisation had to be considered.

Such full studies showed, as with other industries, that there were problems due to the past growth and consolidation of the functional departments. In some large companies, multiple communications systems had developed between the departments ashore and the related departments in ships, and there was often a lack of focus on the product, i.e. the ship. This development had occurred in other industries where functional departments grow to such an extent that they become self-contained organisations and in consequence were inclined to become self-centred.

Some shipping companies solved the problem of focus by introducing a co-ordinator with line or executive authority between the ships and the functional departments. Others, following a relatively new management philosophy, adopted a more complex arrangement known as a matrix organisation or dual accountability organisation structure. Against the background of tradition this is not easy to achieve in any industry and shipping was no exception to this.

The matrix organisation not only lent itself towards co-ordination of effort towards the ships but also assisted in the movement towards delegation of more accountability to the ships staff, that is, towards a form of decentralisation. It was believed that by transferring as much decision making as possible to the ships, more efficiency could be achieved through greater job satisfaction amongst the sea staff. It was hoped that this would also result in a reduction in the wastage or high turnover of sea staff. Importantly there would be less need for supervision from ashore and thus there could be a reduction in the shore staff and their associated costs.

Difficulties were foreseen in providing sufficient information for the managers in the ships to be able to make adequate decisions, but these were solved by careful analysis of information requirements. A major problem lay in the stability or permanence of the crew because of the uncertainty as to the length of time they would be associated with a particular ship. A four or six months period was not considered long enough to develop the requirement interest and a minimum period of two years was considered by many to be the least acceptable. It was also difficult to reconcile these organisational requirements with some new leave allowances of up to 50 percent of time served, and service in ships restricted to four to six months. It is noteworthy that these improvements in leave and reduced sea service had been brought about to encourage men to stay at sea and were now found difficult to reconcile with new concepts which had similar aims.

Additionally there were difficulties in re-education of the shore and sea staff. Both needed to be taught their new roles and associated authorities and responsibilities. Resistance to change had to be overcome and a complete revision of control systems was usually necessary.

While these changes were going on, some continued to manage ships in the traditional ways. Others gradually adopted some of the cost saving ideas explored and practiced by the British, North European and Scandinavian countries. Unfortunately such adoptions were usually carried out without the necessary studies of ships, equipment, systems and organisation, resulting in less cost saving than had been anticipated. For example, although it had been established that ships could be operated by reduced crews the adoption of this idea without the necessary support of related equipment, methods, organisation, etc. often resulted in a reduction in the maintenance and safety in the ships.

In time, others may have to seek new methods and organisations like the older maritime nations for the reasons given. For the present it can be said that there are two distinct styles of management of ships; the traditional, centralised organisations; and the new simulated decentralised organisations with a number of variations in both. These will be considered in detail in Chapter Three.

Because of the differences in management, responsibility, and accountability, associated with these two types of organisation, they will be dealt with separately as necessary throughout this book.

The difference between ship and other management

In general, everything said about management at the beginning of the chapter applies to shipping. In life there is very little original thought and in the same way there are very few original situations found in the management of business and other organisations.

In the very early days of the events which were to lead to the changes in British, North European and Scandinavian shipping, many considered that ship management was different and could not be compared with other types of management. In this they were typical of other traditional managements which had found themselves in a similar position.

But with time, education and experience, there came recognition that management skills, tools, and disciplines, could be applied to and were needed in shipping, as much as in any other industry or organisation. Of course each type of organisation has its own terminology, staff titles, rules, objectives etc., but essentially management theories and practice can be applied to them all. Even the navies of the major maritime nations have long recognised they too have managerial problems which need managerial solutions. They too have employed management experts to seek ways to improve their efficiency and to deal with the problems of staff wastage.

Improved communications nullified the argument that a ship moving around was different to an outpost of a large international organisation. Its relationship to the head office was no different to the subsidiary office of a large company in the next town. Central control was already well established, but a form of decentralisation was possible. There is only one problem in the implemention of a form of decentralisation in ships which singles out the shipping industry from others. That is the stability of the crew. Decentralisation with all that it entails can only work with stable staff, particularly at the senior levels. Ways can be found to decentralise ships but they are not easy and in themselves may increase the costs initially, although there is potential for a decrease in the long term.

Chapter Two

Objectives, functions and restraints

Objectives and functions

It is important to remember that ship management in itself is only part of the whole shipping company. The overall direction and associated long term planning of the company, is a function of the senior management embodied in the board of directors, the management or executive committee or similar body, or even just one person. The prime objective of such a body is to maintain a profitable enterprise. This is achieved, hopefully, through decisions on the type of shipping business to engage in, the type of ships required and how and when to buy or sell them or lay them up. Associated with this will be decisions on the type and amount of financing required, the amount of insurance cover and the type of ship and operational management required.

The short term objectives of the company are, essentially, to keep the ships operational and at the same time on target with the short term plans which link with those of the long term. In many industries these are achieved through functional departments and managers commonly described as:

Production — the production of goods and services.

Sales and marketing — selling the product.

Services — support for the production and sales, such as administration, accounts etc.

Research and development — seeking new products or business.

In shipping, Production relates to Ship Management, while Sales and Marketing relate to Ship Operations and Chartering. The Services are usually utilised by both, while Research and Development draws on the knowledge and experience of the operators and ship managers to carry out their task of seeking new business and suitable ship design and equipment.

The principal objective of operations is to keep the ships gainfully employed while that of ship management is to ensure that ships perform to the operator's requirements, i.e. to ensure they are properly crewed, maintained, supplied, and insured, and most importantly — available.

To achieve their objectives ship managers must function like any other managers: they must plan, control and organise and be flexible. This involves decision making and delegation of work and responsibility.

The organisation of the ship management sector of the shipping company to achieve its objectives will be considered in depth in Chapter Three. The planning, forecasting and control functions of ship management are described in *Running Costs*, as is the evolvement of the basic functional departments ashore and their related departments in ships, i.e.

Crew
Technical
Supplies
Insurance

supported by Administration and Accounts.

As will be seen in Chapter Three, some shipping companies now consider Supplies and Crew to be services in support of small ship management groups. In the same way, others group spare gear and supplies into one purchasing department, but have a separate catering or victualling department. But these are only organisational matters and do not reduce the importance of those functions.

New concepts of ship management organisation and arrangements of authority associated with decentralisation have been mentioned in Chapter One. Regardless of where responsibility for decisions and expenditures lies, or whether the managers are considered to be line or staff (i.e. executive or functional), the shore management also has a support function towards the ship and the ships have a function, ultimately, to produce the required performance.

There are also fringe functions of ship management as in most major industries, such as assisting with industrial forums, committees, union negotiating, etc. The number of activities associated with these industrial functions are usually a reflection of the size of the industry and the involvement of institutions in such matters as the supply of crews and safety.

Restraints

Those involved in the management of ships have restraints placed upon them and they must take account of these in carrying out their functions. As with all management, it can be said that they are restrained on four sides as shown in Diagram 1.

Considering these four "boundaries":

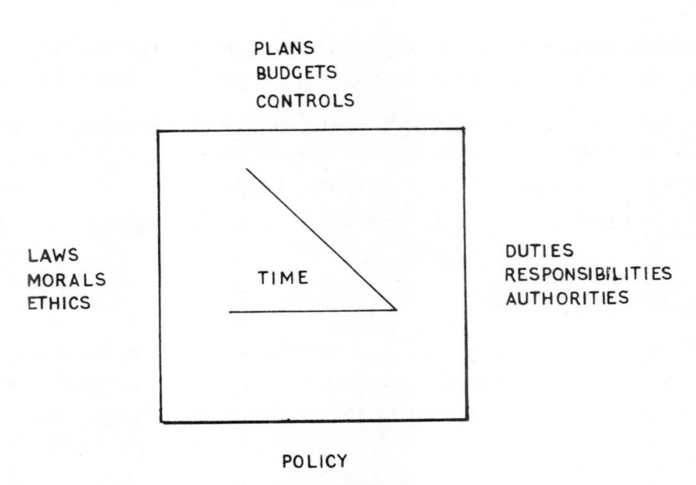

Diagram 1. Diagram of Restraints

Laws and ethics

The principal laws affecting ship managers are the Shipping, Safety, Employment, Environment and Tax laws of the country where the ships are registered. Where appropriate, they must comply with the laws of countries which have adopted international conventions in advance of their own country, or taken unilateral action in shipping matters, such as Australia with its cargo gear regulations and the USA with its anti-pollution and navigation rules.

They are also affected by the shipping and employment laws of other countries applicable to crews, if their country of engagement is different to that of the country of registry of the ship. Similarly they must comply with the rules of any industrial federations to which they may belong and to any union or other agreements.

They may also be restrained by the actions of organisations not directly associated with their crew, such as the International Seamen's Federation (ITF), regardless of its legality in their particular situation.

They must also ensure that the rules of the societies which classify their ships for insurance purposes are adhered to.

Morals (or Ethics) are not easy to define: each business has its own ethical behaviour. As time goes on this becomes known within the industry and related institutional world as well as within the company. Some companies are respected for their moral standards, while others attract disrespect and caution from those who have to deal with or work for them.

A shipping company with high moral standards will often indicate its attitudes in its policies and regulations. Where they are not written they will be well known within the company. While the restraints resulting from these attitudes may cause difficulties for managers at times, in general they are accepted because they are known to be right.

A shipping company with less high standards poses much greater problems for a professional manager. This is particularly so when the line between cost saving and safety is narrow. In this he may be expected to take risks but may not be given support if something goes wrong.

Plans, budgets and controls

Plans and budgets have been covered in some depth in *Running Costs*. Regardless of whether the managers have produced the plans and budgets or had them imposed upon them, once they have been approved at a senior level, they should be adhered to as far as possible. It is noteworthy that if the estimates of plans and associated costs of managing the ships have been produced by the managers concerned, and accepted, they will feel less restrained by these items, because of their personal involvement and commitment.

Keeping the plans and budgets on target requires the necessary discipline of controls. Although at times controls may seem onerous, they are a necessary part of management. Like plans and budgets, restraint decreases in direct proportion to the involvement of the managers concerned, i.e. if the manager organises the controls himself he will be much less restrained by them.

Duties, responsibilities and authorities

Every manager has these to a greater or lesser degree. All too often the responsibilities are not matched by the authorities and this is a common restraint on a manager's effectiveness. However, if the duties, responsibilities and authorities are properly defined and understood, they can in fact assist the manager in his work. It is when they are not clear that his work becomes difficult, i.e. there is a restraint in not knowing what is expected of him or her.

Policy

Policies made by senior management must be adhered to. They usually cover a wide range of matters such that staff do not always recognise them as such, as in the case of the working hours of the company and staff holidays.

In some areas they can assist management by providing rulings, for instance in the case of wives sailing with husbands in ships. Once such a policy is made it avoids recurring consideration of the matter. In such a case the decision is made for the managers and in time the sea staff only apply for what they know to be allowed.

Policy also covers a number of major matters such as the Company's regulations, the commencement and cessation of the Company's financial year, the appointment of Lawyers and Auditors, and Staff expenditure limits. Broad statements of intent as to the long and short term objectives of the company are also matters on which policies can be made and this will be referred to in Chapter Four.

The type of crews, union recognition and involvement with institutions are also matters on which policies can be made. As with plans and budgets, the more staff are involved in formulating the policies, the less restrained they will feel about them.

Another factor

There is also a third dimension, "Time". A manager can only do what he and his staff have time for. If the policies, plans and budgets take proper account of the staff available it is unlikely that time in itself will be a restraint, although priorities and emergencies may require temporary adjustment to work plans. But if the plants, etc. are not properly constructed, and there are insufficient staff with appropriate skills, the manager will undoubtedly be constrained by time in what he can do effectively.

Finally there is an important function of ship management that only applies to organisations which operate continuously. Shipping is a twenty-four hours a day, three hundred and sixty-five days a year industry and although some of the staff involved need not be available at weekends and holidays, sufficient key staff must, of necessity, be available at all times ashore.

All these restraints have great influence on the functions of ship management and managers should be guided accordingly.

Chapter Three

Organising ship management

"The beginning of administrative wisdom is the awareness that there is no one optimum type of management system"

Tom Burns

Organisations in general

Structure: Organisations should be planned and arranged for "results" and thus efficiency. An essential objective in arranging any organisations is the smooth flow of communications, i.e. information, decisions, advice, etc. between the parts without which management cannot be properly exercised. The way in which the communications flow will depend to a large degree upon the type of organisation, i.e. whether it is centralised or decentralised. In organisations where the control is centralised the information about the work done flows from the subsidiaries or satellites towards the center, while instructions flow out. In decentralised organisations, where the subsidiaries control themselves to a large degree there is a change in the flow, with advice and information for the subsidiaries being passed out from the center and the subsidiaries advising the center of the actions they have taken. Alternatively information may be exchanged so that joint decisions can be made.

In both types of organisation there is also a need for communications between the functional departments, the "service" departments, the product center and the rest of the organisation.

Whichever the direction of the communications and the type, they must be logical and thus well planned, otherwise confusion will reign. Simplicity should be the key word. It is very easy to create a complex organisation with complicated lines of communication. An efficient organisation is not easy to create *and* maintain.

Before considering shipping company organisations, it is important to bear in mind one very important factor about organisation charts. It is that the arrangement of the departments, and particularly the people within them, are drawn to show the relationship of one to another, and in a further development which will be considered in Chapter Four, the flow of communications. They do not necessarily indicate seniority of departments or people within them.

Shipping company organisations

As mentioned in Chapter One, there have been considerable changes in shipping company organisations since the 1960s, but there are still many whose organisations closely resemble those in existence in the 1930s, while others are at various stages between.

It is therefore preferable to examine each of the basic types of organisation. In doing so it should be appreciated that there are many variations to these basic themes, in the same way as there are often different names given to the same function. For example "Crew" is often called Personnel today in line with other industries. Supplies, Stores, and Spare Gear are often grouped together as "Purchasing". In themselves the names are not important, providing all concerned know the purpose of each department or section of the organisation and the authorities and responsibilities of the staff within them.

The 1930s standard organisation

This was a development of the end of sail organisation which had looked something like this:

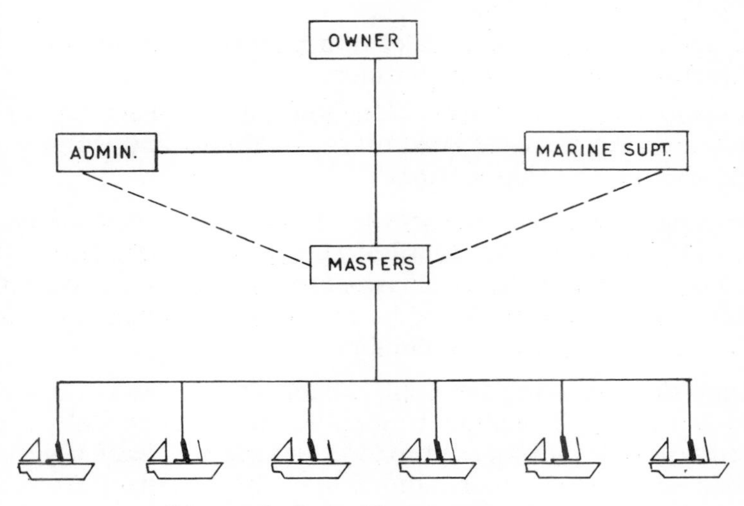

Diagram 2. End of Sail Organisation

The introduction of steam and the requirement of additional technical knowledge, plus improved communications resulted in an organisation in the 1930s which looked something like this:

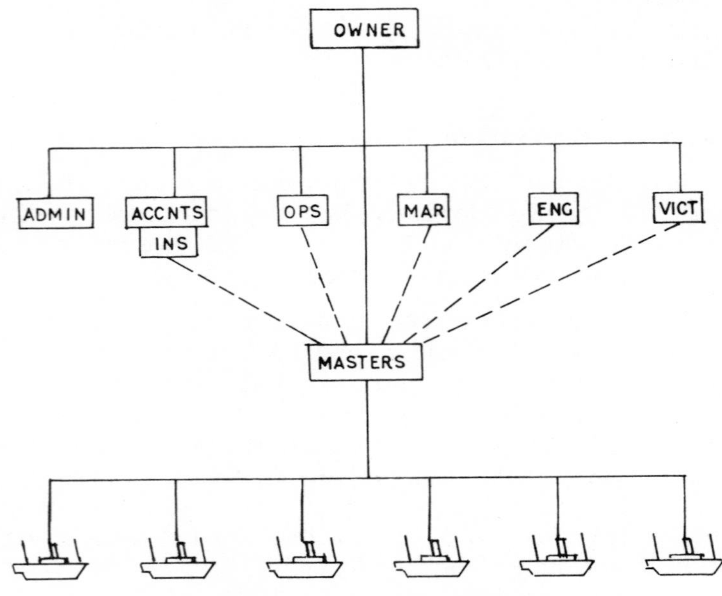

Diagram 3. 1930s Organisation

The following developments from the first chart are worth noting:

The separate Engineer Superintendent's Department.

The addition of a Victualling Superintendent whose prime concern was the supply of food to Ships.

The recognition of the importance of accounts and insurance, although still as sections rather than departments.

The development of the "Sales" function, i.e. the chartering or liner departments ashore, as the head office could now find world-wide business without the ship masters.

Particularly significant was the fact that at this time most communications were channelled through the Board of Directors (BOD). This had a co-ordinating effect and also emphasised the direct "line" association of the Master to the BOD. It should be noted that the communications in those times were usually few in number.

Today some shipping companies still organise themselves in this way. They may not use a Victualling or Supplies Department, delegating full responsibility for all purchases abroad to the Master. Similarly responsibility for minor repairs abroad may also be delegated to the Master and Chief Engineer. The prime role of the Marine and Engineer

Superintendents, or Port Captains and Engineers, in such an organisation is supervision, trouble shooting, major repair and maintenance work, and advising the sea staff of new legislation and technical information.

This type of organisation may still be well suited to an owner with unsophisticated ships and officers and crews with only basic training. The very sophisticated techniques of modern ship management cannot be used in such an organisation to gain efficiency, because there is neither the equipment or trained staff in the ships or ashore to cope with them. However, some techniques can be adopted to advantage such as planned maintenance or spare gear control. They can be installed by outside experts and can be maintained, providing there is continuity of the key staff and that they understand the systems.

The choice of appropriate Masters and Chief Engineers for the ships of such a company is vital to the owner. This is particularly so if there are no defined authorities, instructions, etc. In such a company the shore organisation is very "lean". Most of the staff have a wide range of responsibilities and communications between the staff if informal, fast and therefore efficient. Major planning and budgeting is done by the shore staff while the Master and Chief Engineer have an important role in keeping them advised of expenditures and requirements (such as Spare Gear) which are not readily available to the ship in most ports.

This organisation is, broadly speaking, centralised, because although the Master is left to make many decisions on his own, he is not fully "accountable" for the management of the ship within the meaning of the word as described in the introduction.

The 1960s functional organisation

The increase of technical knowledge, coupled with greatly improved communications and control techniques as described in Chapter One, resulted in a growth in the size and development of the functional departments of many older shipping companies. This resulted in an organisation which looked something like the diagram on page 16.

It is easy to understand how such an organisation developed, particularly in the older, larger, shipping companies. New technology, changes in shipping laws affecting the construction and equipment of ships, and greater emphasis on safety and anti-pollution measures necessitated more staff to deal with modifying existing ships, providing specifications for new ships, and keeping sea staff informed. Similarly the studies of work in ships and the resultant new systems, needed extra staff and time for their development.

As mentioned in Chapter One, the problem with this type of organisation was the strength of the functional departments. The system of communicating through the BOD had gradually disappeared and the Departments were communicating directly to the ship with regard to

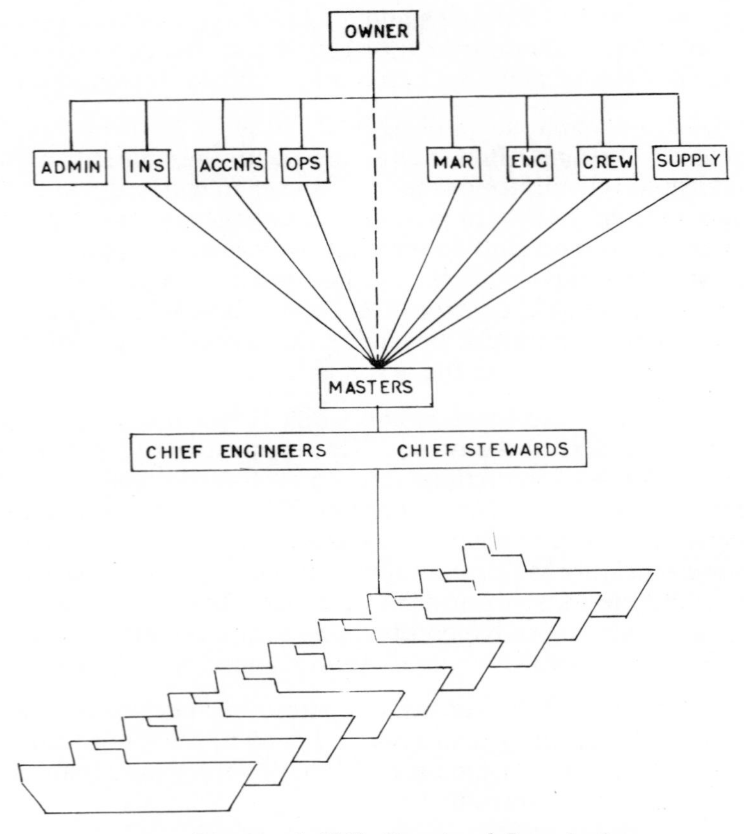

Diagram 4. 1960s Functional Organisation

their particular needs. In most cases the communications were still through the Master, but there was also a tendency to write directly to the corresponding department in the ship. This, together with the development of "new" management techniques requiring budgeting and planning on a departmental basis, further consolidated the "functional" aspects of the organisation ashore and in the ships. It also created an anomaly in the ships, where in the new systems of working, the staff were required to work as a team and discard the departmental barriers.

Officially there was no one to co-ordinate the various functions directly associated with the ship, although the operations departments of many companies often acted in this role when necessary. Recognising this need some companies appointed a co-ordinator or fleet manager, and this did produce some focus on the ship as a whole. The resultant organisation looked something like Diagram 5 on page 17.

Although recognised as a step in the right direction the role of the functional departments remained the same and thus the co-ordinator had little real authority, although appearing to be in a key position on the

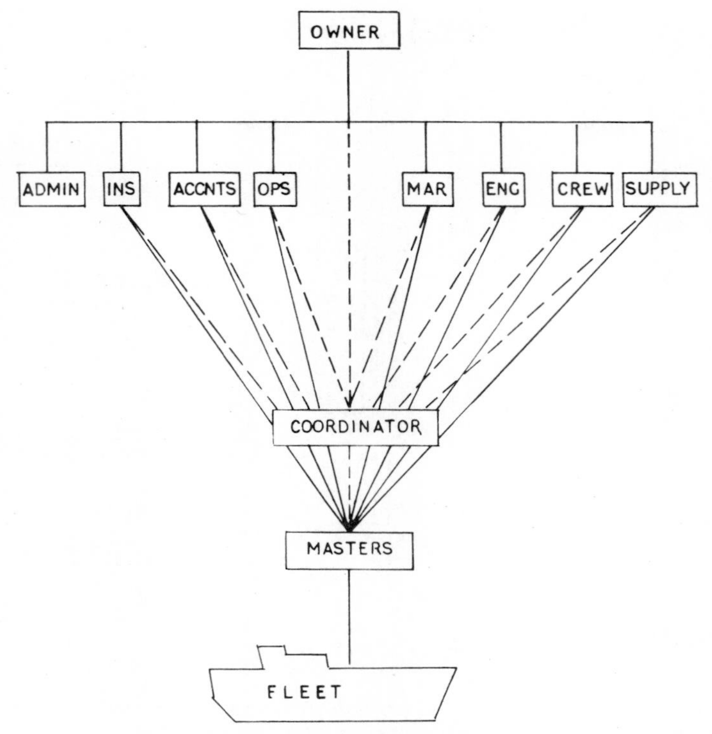

Diagram 5. Introduction of the Coordinator

chart. In the search for an improved organisation which would rectify this situation, some shipping companies turned towards the matrix organisation which was already in operation in some large companies in other industries.

The 1970s matrix organisation

This was system of dual accountability of line and functional managers. It provided greater focus on the product or ship and yet allowed accountability of both the functional and Line managers. The organisation for a shipping company of about thirty ships as shown in Diagram 6 (p. 18).

Within the Fleet Management, the functional managers (Crew, Technical and Supplies), were responsible for ensuring that their individual functions were carried out in all the thirty ships, while the Fleet managers were responsible for ensuring that *all* the functions were carried out in each one of their fleets of ten ships.

This required a considerable amount of liaison between the fleet managers and the functional managers, e.g. the functional managers were responsible for their total departmental budgets and yet had to

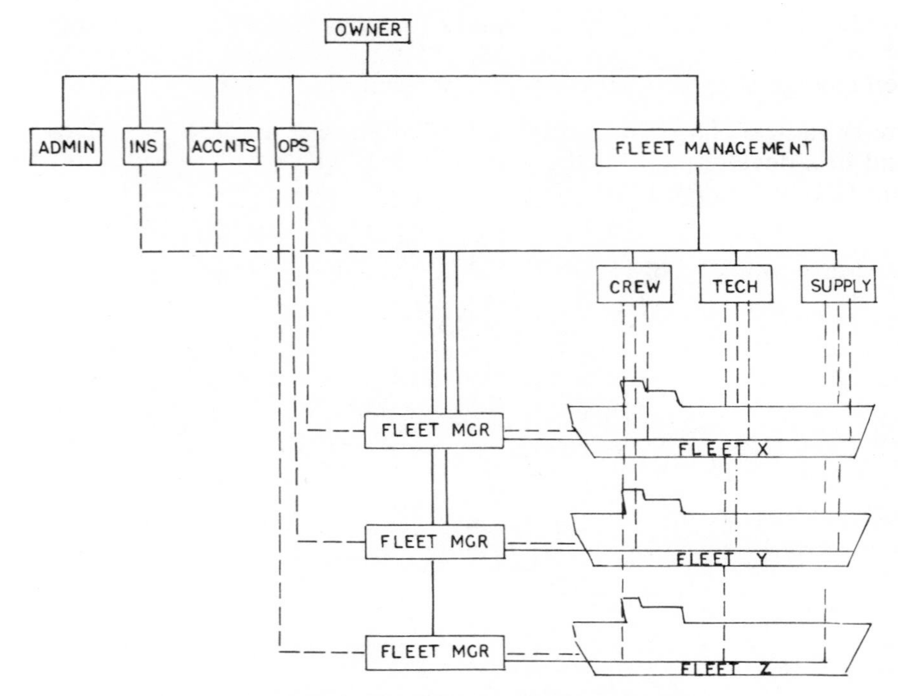

Diagram 6. The 1970s Matrix Organisation

agree their ship department budgets with the individual fleet managers. Extraordinary expenditures outside of budget had to be agreed between them and explanations for budget variances had to be provided for their mutual chief – the Head of Fleets or similar title.

Additionally the Fleet managers had to work closely with the operators to ensure that the ships were available as required, and that periods out of service were co-ordinated at optimum times and places.

Essentially there were two problems with this type of organisation:

The number of ships allotted to each Fleet manager were too many for him to have a close working knowledge of them.

Although providing a focus for co-ordination, the Fleet manager's position still lacked real authority and effectivenesss, because of the continued size and power of the functional departments. He managed by monitoring and questioning the actions of the functional managers. If he did not agree with their actions he could challenge them, but if agreement could not be reached the matter had to be referred to their mutual chief.

As already mentioned, resistance to change was probably a factor in preventing this system from working as intended. The residue of past systems in the functional departments still had a great influence on staff attitudes, such that they often resisted the new, and tried to continue in their old ways as far as possible. This is not an unusual phenomenon when change is forced upon people, particularly in industry.

A more radical change was required to emphasise the new way. This was found in a development of the matrix system which can be called the "Ship Group Organisation".

The Ship Group Organisation

It was generally agreed that the role of the co-ordinator was essential, but it was found that small groups of ships of about three or four, were more effectively co-ordinated or controlled than larger groups. To fill the role for this task the position of "Ship Group Manager", "Project Leader" or just "Ship Manager" was created with overall executive or line accountability for the ships in his group.

The organisation which resulted looked like this in a shipping company of the 1960s:

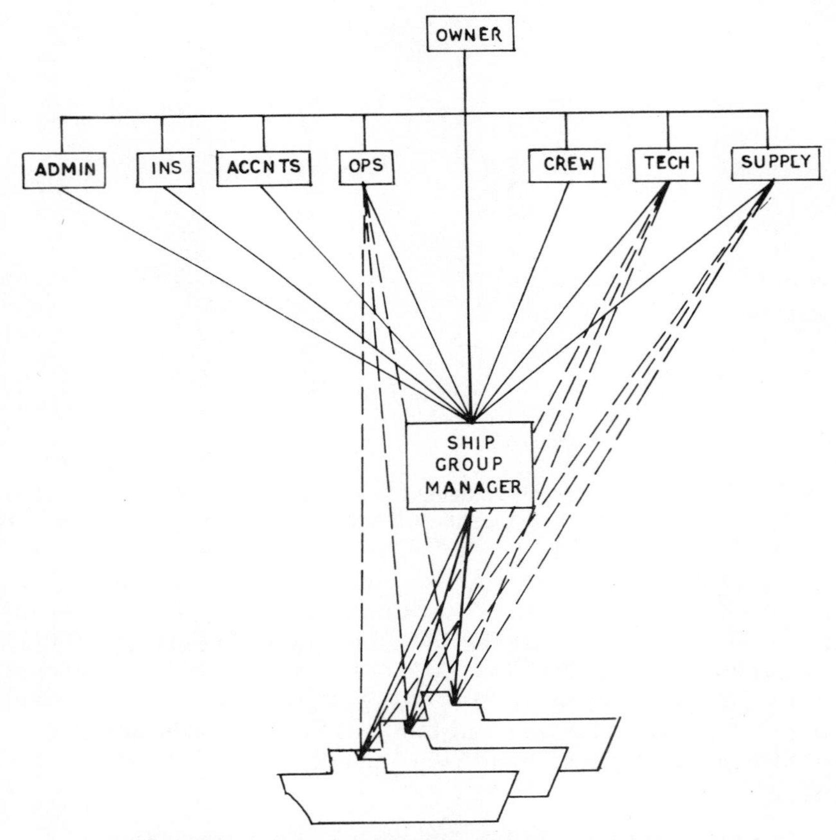

Diagram 7. Introduction of the Ship Group Manager
(see note to Diagram 8)

When combined with the matrix organisation the resultant organisation looked like this:

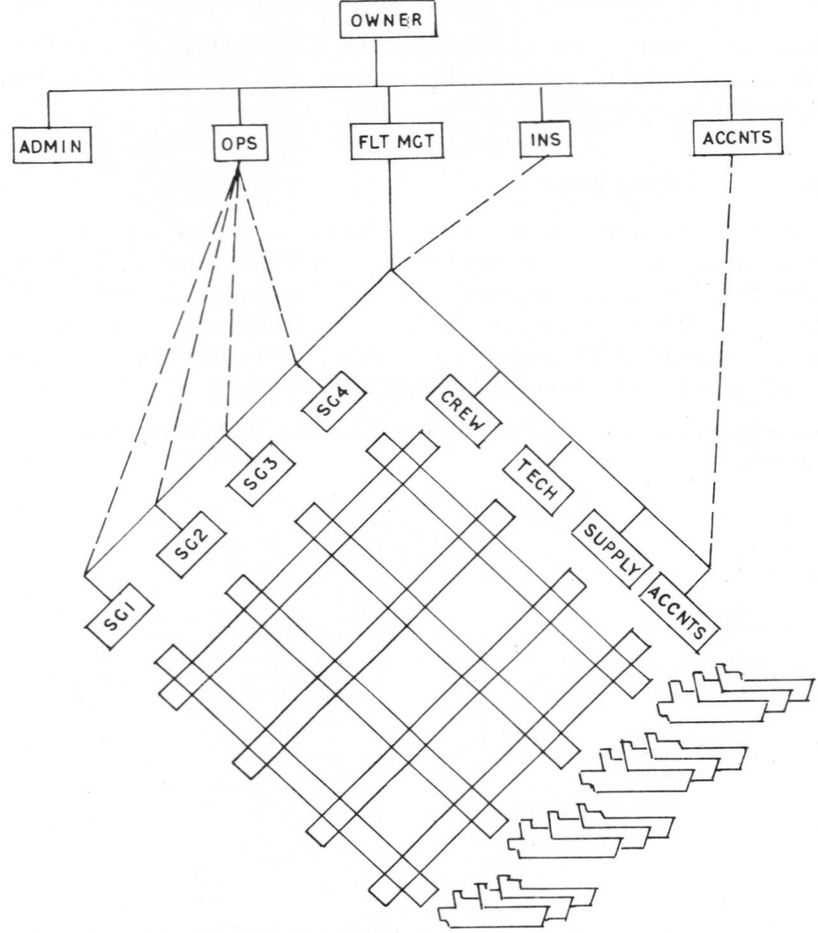

Diagram 8. Ship Group — Matrix Organisation
NOTE:
Diagrams 7 and 8 are developments of figures from *Emerging Organisational Values in Shipping* by M. H. Smith and J. Roggema, which were adaptations from figures from *Drift Av Skip* by Duckert and Kevin.

At first the Ship Group Manager operated alone (apart from secretarial services), calling on the functional departments of Crew, Technical, and Supplies, for support for the ships of his group, or ensuring that support was provided. In time it was found that as so many important line decisions involved technical matters, a technical expert should be added to his staff.

There was thus a movement of technical staff from the "functional" to the "line" side of the dual accountability organisation structure. At the same time there was a tendency to merge the Crew and Supplies departments into one "Services" department to support the fleet as a whole. Some companies also set up a fleet dry dock department to deal with major repairs and modifications. Another central technical development department attended to the issue of new technical information (marine, engineering and electronic), and technical developments in the fleet as a whole.

The organisation which evolved looked something like this:

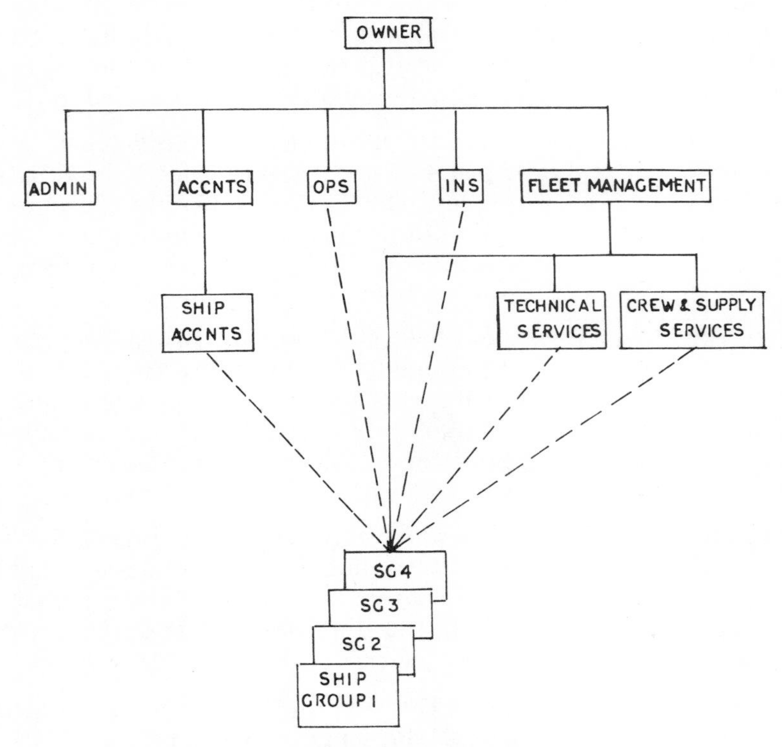

Diagram 9. Ship Group Development

Simulated decentralisation

Reference has already been made in Chapter One to the studies of work and organisations in ships and the Shore offices. The matrix organisation re-emphasised the direct line association or chain of command between the BOD and the Ship Master, through the Head of the Fleet and the fleet manager. As a result of the early studies the sea staff were better informed, and to some degree more involved, but the shipping company organisation was still a centralised one.

The studies went on as costs continued to rise and still more efficiency was sought. Every company that engaged in this search either employed outside experts to assist, and/or used their own "in house" expertise if they were part of a group large enough to employ their own experts. In some cases the expertise was provided by the "institutions". In almost every case there was full consultation with all the unions concerned. Where appropriate, government departments and industrial federations were also involved.

This joint approach to seeking answers to problems affecting individual shipowners and the industry as a whole was found to be beneficial to all. The knowledge gained was not confined to the industry of one country but was generally available to those seeking the same solutions. This was particularly so between the North European maritime nations of Norway, Sweden, Germany, Denmark, Holland and the United Kingdom.

In 1974 Jebsens (UK) Ltd., a British subsidiary of a Norwegian company, approached the British Tavistock Institute for help in its search for a solution to its crew problems. The Tavistock Institute is renowned for research that concerns itself with bringing about change and solving problems in the systems in which people live, and their experience brought a fresh approach to the problems.

Shortly afterwards, in 1975, the General Council of British Shipping embarked upon its "Sealife Programme" supported by members of the British Maritime Industry. For the next four years this programme focussed upon the manpower supply problem in the British Shipping Industry, using research methods for measuring activities in ships developed by the German Flensburg Project.

The programme included studies on recruitment, selection, induction, training, shipboard management, crew stability and the design of ship superstructures and engine rooms. A number of well known British shipowners participated actively in the studies including Panocean ANCO and Denholm Ship Management.

The conclusions reached by those carrying out these studies reinforced earlier indications from the British shipping industry that there was potential for greater utilisation of ships crews, particularly the senior officers, by greater delegation of authority. Similar conclusions were being reached by the shipping industries in Germany, Holland, Denmark and Norway at the same time.

It was recognised that full decentralisation, as defined in the introduction, was impossible to achieve because the trade for ships could not be sought by the sea staff, and major technical work and purchasing could not be satisfactorily negotiated and arranged by them, because of their relative isolation. But much could be delegated and an organisation of the type known in other industries as "simulated decentralisation" (SDC), was introduced by a few companies of which the Danish shipping company DFDS was one of the most progressive.

In this system of organisation the shipboard management team is accountable for all areas of costs and earnings which they can feasibly control, given the necessary support, information, and advice. They can thus "manage" within the constraints placed upon all managers as described in Chapter Two. In matters over which they do not have full control, they can provide the shore staff with the necessary information and comments for decisions to be made jointly or separately. Ashore, fewer staff are required and attention is focussed on ships rather than on departments, although experts are still required to give advice and make decisions on matters outside the scope of the shipboard management teams.

It must be emphasised that in shipping today, simulated decentralisation is a relatively sophisticated and long term system. It requires highly qualified, developed, flexible and motivated people, both at sea and ashore. A company changing from a traditional organisation to simulated decentralisation requires considerable commitment to the system by all the staff, particularly senior and middle management.

The re-organisation needs to be very carefully planned, particularly in regard to the flow of information and the definition of authority. This planning is time consuming and initially expensive, as staff at all levels, both ashore and in the ships, need to have varying degrees of training to assist them in their new roles. Even a new company starting with this type of organisation needs to ensure that staff recruited are familiar with the roles required of them.

The SDC system blends well with the Ship Group System. The only change that is required is one of emphasis in the role of the Ship Group Manager. He still holds "line" authority, but his task has a larger co-ordinating component than that of directing, as in the ship group system.

Conclusions on the types of organisation

Apart from the restraints of institutions referred to in Chapter Two, shipowners do have a choice of organisation, and that choice should reflect the type of ship, the type of crew, and its training and abilities, and the size of the company and its planned future.

A modern, technically sophisticated ship, demands a highly trained crew. Similarly, highly trained crews desire responsibility and involvement. For a shipowner with a number of such ships and crews and a reasonably foreseeable future, the simulated decentralised management organisation seems preferable to a centralised one, because of the mutual benefits to employer and employees, i.e. a smaller shore staff and greater job satisfaction and thus "motivation", for the sea staff.

It also strengthens the shore management, as the delegation of many of the short term decisions to the sea staff leaves them free to either move into "line" management or make greater use of their functional skills. When legislation allows greater flexibility in the roles of staff in ships the SDC organisation will be further enhanced.

For shipowners with sophisticated ships and highly trained crews, who do not feel inclined or be able to delegate as much responsibility to the sea staff as is required by SDC, the developed Ship Group concept is a viable alternative. The ship group team still emphasises the focus on the ship unit and the inclusion of technical expertise in the small team strengthens the line authority of its members. As with the SDC, the functional departments of crew and supplies can be combined into a services department. Similarly specialist technical departments for longer term projects can be provided as required or consultants utilised. In such an organisation it is usual to include a supervisor in the Ship Group team for the greater number of ship visits required by this more centralised type of management.

The ships' staff can be involved in costs and controls and in a few areas such as overtime can be made directly accountable for targets they set. They too will work as a team relating to the Ship Group team.

For a company with unsophisticated ships, with a policy of employing casual labour at all levels, there is little choice but central control along the lines of the 1930s model. An alternative to this is a smaller central staff and the appointment of permanent and reliable Masters and Chief Engineers, who make the majority of short term repair and purchase decisions themselves during the voyage. It can be argued that the costs of wrong purchases at the wrong ports is less than the costs of maintaining a larger shore staff and systems to monitor and control those costs.

For major repairs, consultants can be used to provide specifications, obtain tenders, choose the repair yard, monitor the repairs and finally approve the costs. Again this will save the costs of maintaining a large technical staff, although someone will have to make arrangements to keep the ship advised on technical matters and to process such matters as spare gear orders and deliveries, and co-ordinate crew changes. For an owner with only a few unsophisticated ships this type of management is very suitable, but its effectiveness depends to a large degree on the two seniors in the ships. As will be seen in Chapter Twelve, regardless of the efficiency of these two, the owner will still need to have then supervised if only to protect himself.

Owners who are able to obtain low cost crews can often remain competitive, although maintaining a management organisation similar to those existing in the 1930s. By adopting some of the philosophies and systems mentioned in this Chapter, they can probably achieve an improvement in their ship efficiency. But as has been said before, people rarely seek efficiency voluntarily and change usually has to be forced upon them by events.

Whatever the organisation, the best results will undoubtedly come from staff who are motivated by personal commitment to their task. Involvement helps, but it is not as effective as the delegation of authority to staff who set and control their own plans of action, and are accountable for the results. The simulated decentralised organisation provides the best solution to date. Compromise organisations are not as good, but if some of the general concepts of delegation and accountability can be included they can help both owner and employee.

Because of the essentially different requirements of staff in a centralised and simulated decentralised management, where appropriate these will be considered separately in Part Three.

Which organisation?

When some ask, "How much will it cost to run a ship?" The answer should be: "It depends on the factors involved". The same reply should be given to anyone asking about a preferable management organisation for a shipping company. Whatever the factors and the decisions taken, in the end it will be the people involved who will make it work, or otherwise. Whichever system or organisation is used, there seems little doubt that motivation is a key factor. This will be considered in Chapter Fourteen.

Supporting the structure

Having decided upon the type of organisation best suited to manage the company's ships, the next step is to decide the number of people and skills required to support the various parts of the organisation. This should not exclude consideration of the use of part time or outside expertise in certain areas if this is more suitable financially and practically.

In larger companies some senior managers will be required to concentrate all their attention on the long term objectives mentioned in Chapter Two, while other managers will be required for the short term, day to day, running of the company. In smaller companies it is probable that they will all be engaged in a mixture of the two.

If these managers are to be able to use their skills to the full they will need to be supported by assistants and administrative staff; bookkeepers, telex operators, filing clerks, typists, secretaries, telephonists, mailing clerks, etc. The number and flexibility of these support staff will also depend on the size and complexity of the organisation. Some may be attached to a particular department, while others may provide a central service such as the telex operators and telephonists. In a small company, funds may not allow for the ideal support for the key managers and they and their assistants may have to undertake some support tasks themselves, e.g. filing their own documents and even typing their own letters and sending their own telex messages. This is one of the difficulties of very small organisations; that expertise, is of necessity, often wasted on mundane but nevertheless essential tasks.

In small companies, a decision is usually made from experience, on the qualifications of the person required to manage a department or project and the support staff required. In large and more formal companies, a written position specification may be prepared for each position in the organisation. This will describe the responsibilities and duties associated with each position and also indicate the age, qualifications and experience of the person required.

The specification may go further, and define the authorities of the position holder in terms of cash expenditure, and may even describe the extent of his authority in terms of "ACT", "ACT and TELL," or just "TELL". An example of this is given at the end of Chapter Thirteen for a master and chief engineer officer. It will be appreciated that in most organisations not all "ACT" and "ACT and TELL" or just "TELL" situations are clearly defined, and managers have to develop a sense of what they can do alone, or should report that they have done, or should ask about before taking any action. Whether or not the authorities are defined, there will always be the case where a junior thinks there is no need to report something and his senior thinks he should. Although not a perfect solution to this age old problem, the position specification with specified authorities is helpful in guiding staff on how to act on major matters.

Eventually the position specification may be changed into a position description, which ideally, should be written by the position holder and agreed by both his senior and junior. It is often surprising how differently the senior and junior view the position compared with the holder, and from this point of view alone it can be a worthwhile exercise. Unfortunately, both the position specification and position description are sophistications, and although little training is required to prepare them, they do absorb time which smaller companies can rarely afford.

In ships there are often legal and institutional restraints such that there is little scope for flexibility in the standard positions. Despite this, the setting down of descriptions can be helpful to staff, particularly when starting in a new position, or when the roles associated with certain positions are altered, e.g. at the change from traditional to general purpose manning.

The examples of descriptions for a Master and Chief Engineer at the end of Chapter Thirteen, illustrate the way in which sea staff positions can be described.

The background and qualifications of the various key staff in the ship management section of the shipping company will be considered in specific chapters of this book.

Organisation and staff in practice

A final point on staff and organisation:

Organisations rarely stay the way they were planned initially. This is not only because of the changing needs of the company, but because of the different capabilities of individuals, particularly managers.

For this reason positions are often combined under curious titles — not so much because the position is logical, but because the man holding the position can do the work.

Part One

CONCLUSION

Essentially the factors involved in managing ships are no different to those involved in any other business.

The objectives are the same as those in any other business.

The functions are essentially the same, although dressed in different names.

The restraints are the same, and like other businesses, are less onerous if the staff are involved in setting those within the company's control.

Whichever way a shipping company is organised, the basic ship management functions must be covered by someone inside or outside the organisation.

There is no one way of organising a shipping company, although there is probably an optimum way for each company, considering the owner's style of management and the staff, ships, equipment, and systems. But of these the staff is vital. It is wasteful and even dangerous to put untrained men with sophisticated systems and equipment. It is also wasteful not to use the potential of highly trained staff.

When considering a shipping company organisation, the total organisation of ships and shore should be studied for maximum efficiency.

New techniques and systems can be adopted piecemeal, but are not as effective as when they are properly integrated into the total organisation. They depend upon the staff who operate them.

Attention should always be focussed on the ship and the company, and not on departments. In concentrating attention on ships, small groups managing a few ships are preferable to large groups managing many ships.

Delegation should always be to the lowest possible level. Responsibilities and authorities of staff should always be clear. Whenever possible sea staff should be accountable for areas they are able to control.

Chapter Four

Regulations and policies

An examination of many shipping companies' instructions to masters and sea staff, would reveal much variance in the way shipowners tell them what they want done or not done, present their requirements, and how they name their requirements. It would also show a considerable variance in the number of instructions, from a single letter to a set of manuals. One can only speculate on how this situation has developed.

The background

It probably started in early shipping companies with a letter of appointment to the master taking over command of a ship. In the letter he was sternly advised in the language of the times, to adhere to the laws of the sea and to take care of the ship of which he had been put in charge and all on board. In view of the historical practices of shipmasters, he was also instructed, under threat of instant dismissal, not to trade for his own account or to accept unofficial payments.

In time the letter became "standard" and by the early part of the twentieth century had developed into a small printed booklet known as the "Company's Regulations". Again one can only speculate on the expansion of the letter into a set of Regulations. Doubtless, groundings and the loss of ships, prompted the inclusion of instructions on such matters as the taking of soundings when approaching land, and later, the use of the Radio Direction Finder.

But legislation regarding the limitation of the shipowner's liability must have affected the instructions also. The owner could only limit his liability if he could show that he had provided a properly managed, seaworthy ship, with a competent crew; and a qualified, experienced master, who clearly understood his requirements as to the safety of the ship, life, and cargo. It was thus very important that all instructions given were clear and so worded that they would stand the test of any legal action.

With experience of such legal actions involving the shipping company or others, new instructions were added from time to time. Similarly, as time went on, unsuccessful claims against insurance highlighted the need for specific steps to be taken at times of accident and damage to ship and cargo. These were also included in the regulations.

Although the management of functional departments referred to in Chapter Two undoubtedly increased the number of instructions issued to ships, the growth in the number of instructions was also due to increased legislation, directly or indirectly associated with the industry. For example, direct industrial legislation may have resulted in instructions on boat and fire drills, while indirect legislation may have resulted in instructions on guard rails around hatches. The introduction of management controls would give birth to instructions on the measurements and reports required, and their frequency.

At first the instructions from the functional departments were usually signed by a director and addressed to the master alone. This was because of the direct link of master to the Board of Directors, but also because it was felt that it was legally essential to communicate in this way. In time this gave way to managers signing such letters on behalf of the company. This is also referred to in Chapter Three in relation to the ship/shore organisation.

By the 1960s a number of changes had gradually taken place in many shipping companies in different maritime countries, but certainly not all companies or all countries. The marine departments of governments were extending their requirements and warnings, and these needed to be passed on to ships. Similarly the institutions were also producing rules, such as joint employers-union agreements, which required the attention and adherence of the Master.

At the same time, management terminology, if not practice, crept into the industry, and sea staff became used to hearing statements such as "the ship is to be well maintained to company's standards". Another significant phrase was "it is the Company's policy".

By the 1970s the effect of the activities of governments, institutions, and management practice, had gained considerable momentum: world concern about shipping safety and pollution from ships had resulted in a number of international conventions, recommendations and codes, which were adopted by the major maritime nations. Some considered the legislation inadequate or so slow in being adopted, that they acted unilaterally: for example, Australia in regard to cargo gear regulations, and the USA in regard to anti-pollution and safe navigation in its coastal waters.

The institutions, such as the International Shipping Federation (ISF), and the International Chamber of Shipping (ICS), also commenced issuing warnings and reports of incidents. Similarly P and I Clubs produced guidelines and warnings on matters affecting the carriage of cargo in ships, and the prevention of accidents to ship and shore personnel. Even outside organisations, such as the World Health Organisation (WHO) and the International Labour Organisation (ILO) affected the industry through their decisions and recommendations.

Coupled with this was the rapid increase of knowledge and experience in the shipping industry: new concepts such as safety and management committees; new equipment such as crude oil washing, satellite navigation, radar, machinery condition monitors; new systems such as planned maintenance, spare gear control and budgeting; and finally new ship types and cargoes. All these required new knowledge, and each and every one of them required at least instructions and advice to the sea staff involved with them, and often special training too.

The hazards of incompetent operation, or misinterpretation of results from some of the new equipment was such, that many governments and classification societies demanded proper operational instructions to be provided in the language of those operating it. For instance, as in the case of crude oil washing equipment, or in simulated training on radar. Yet another example, was the USA requirement that their special bunkering procedures be followed when ships were taking or transferring fuel in their waters.

The outcome of all this varied from company to company depending upon their attitude. Of course, when procedures and rules were required by legislation, these were placed on board. But otherwise, it remained a question of whether or not the shipowner believed in telling his shipmasters his requirements categorically, or whether having placed the master in command, he relied upon his training and experience and gave him the minimum of instruction.

For those companies which believed in instructing and guiding their shipmasters, the new "knowledge" had resulted in a significant change from the slim regulation booklet of fifty years or so earlier. For some, large sets of manuals had been developed and it was possible to find all the following Manual titles in one ship:

Company Regulations
Company Instructions
Company Policy
Safety Instructions
Technical Manual
Supplies Manual
Crew/Personnel Manual
Cargo Manual
Port Information Manual
Plus Operating Manuals for systems such as spare gear control, planned maintenance and stock levels.

It should however be noted, that in those companies which believed in instructing their sea staff fully, there seems to have been a distinct relationship between the amount of instruction given and the size of the shore base, particularly the size of the functional departments.

There also seems to have been differences of opinion as to what was a regulation, an instruction, a rule or a policy. But despite the different arrangements and titles, there was considerable similarity of content in the manuals of many shipowners who endeavoured to keep abreast of developments in the industry. It is interesting to note that the British Royal Navy's Queen's Regulations and associated manuals show a similar process on the basic essentials of running ships aside from matters affecting their special purpose, i.e. trade or defence. Their development probably resulted from similar reasons to those of private shipowners. But navies are fortunate in that they have all their ships under "one management", whereas merchant shipping does not, and thus there was and still is, little standardisation.

The issue of codes of practice by governments such as the UK Department of Trade's "Code of Safe Working Practice"; and institutions, such as the International Chamber of Shipping's "Guide to Tanker Safety", were a particularly significant development. Previously, governments and institutions had usually only concerned themselves with the interpretation of the law into rules, with regard to the equipment and inspection of ships, and the syllabus for certificates of competency. Regulating actual practice had been left largely to the shipowner and the "ordinary practice of seamen". Generally this was an unsatisfactory state of affairs and although many shipowners regulated themselves very well, there were some who did not consider such sophistication necessary. More was required than the issue of new laws by governments, and so the Codes of Practice came into being. In some companies they filled a gap. In others, where adequate instructions had been developed already, duplication was the outcome.

How much regulation by the shipowner?

Until the 1950s the system of work in ships had been almost universal. Watch systems were the same, the duties of officers of each rank were very similar, and most importantly, senior sea staff of shipping companies had usually been with the same company since their first day at sea. Thus they were fully conversant with the company's requirements and their duties and responsibilities. Today with changing staff, often of different nationalities and professional experience, and often provided by a central "pool", the situation is quite different.

As has been shown, merchant shipping is now a highly complex "detail" industry. It has great potential for danger and damage to life and property. It is therefore essential that those who sail in ships are adequately instructed and regulated so that standards of operation can be set and maintained, and staff know what is expected of them, particularly in emergencies.

Essentially instructions fall into two categories:

Technical or operational: concerned with the technical running or operation of the ship.

Administrative: concerned with the company's system of management.

"Technical" covers all those items such as lookouts in fog, soundings approaching land, fire hazard in machinery spaces, watch keeping, safety, pollution prevention, etc. An examination of the manuals of shipping companies who do believe in instructing ships in some detail, would probably reveal much similarity of content in this area and a potential for standardisation. As already noted, some of the instructions under this category have been standardised by governments and institutions.

"Administrative" covers such items as frequency and methods of communication with the head office, procedures following accidents, responsibilities of officers in regard to operational matters, wives and visitors in ships, noting protest, emergency communications, salvage, cash, etc.

Obviously there are grey areas between these categories, and some operational instructions may be considered a necessity by one company and thought to be so traditional and accepted by another that they are not required.

There is much that could yet be standardised for the benefit of the industry either by country or internationally, but such matters take a long time to arrange. Thus for the present, one must consider what can be done with a limited amount of standardisation in conjunction with existing laws and individual company regulations.

For any shipowner considering the issue of new or revised instructions to ships the following are suggested guidelines.

Avoid duplication: If the matter on which it is considered the staff need instruction is adequately covered in an official or approved industrial publication, use that publication. Refer to it in the regulations and ensure that copies of it are placed in the ship. If it is felt that additional instruction on the subject matter is required, include it in the instructions but stress that is supplementary to the publication.

Avoid obsolete matter: Some instructions become outdated through time and yet are retained. The relevance of all instructions should be considered regularly.

Keep instructions to a minimum: All instructions should be thoroughly examined before issue to consider whether they are really necessary and if so whether the regulation is as concise and clear as possible.

Issue instructions which assist rather than demand obedience: The prime purpose of instructions should be to provide a point of reference for sea staff to turn to if in doubt of the action they should take.

Avoid the automatic issue of all government or institutional publications. It is not enough for technical shore staff to just pass on every piece of information or warning they receive to the sea staff. They have a responsibility to examine such information carefully and only pass on that which is relevant, adding their comments as appropriate.

Organising the instructions

The title: An examination of the definition of the words "regulation", "instruction", "order", "rule" and "procedure" in a good dictionary shows a similarity of meaning and for persons not involved in the legal subtleties of such meanings, they all seem to have a common thread, i.e. to regulate and guide.

Policy has a somewhat similar meaning and when used in the context of, for example, "it is the company's policy that the Master shall be in charge of the ship when approaching port", takes on the force of a regulation. However, policy from a business management point of view, has a somewhat different meaning and this will be considered separately at the end of the Chapter.

It is therefore proposed that all instructions be called "Regulations", to avoid confusion.

Arranging the regulations: Ideally the regulations should be arranged in a loose leaf ring binder in order to allow amendments from time to time, although much of the content will probably be of a permanent nature.

Each page should show the date of issue, but pages need not be numbered if each section is numbered, for example, if the first section is on Navigation, then each subsection of that section can be given a subcode, as follows:

2.00 NAVIGATION
2.01 APPROACHING LAND
2.02 CHECKING RADIO DIRECTION FINDER
2.03 COMPASS ADJUSTMENT

and so on. This allows additions to be made to the various sections without disturbing any page numbers. Finally, the manual should have an adequate index and an amendment page on which amendments can be recorded.

Sufficient copies should be placed in each ship so that at least the senior officers each have a copy.

The contents

In the first place it is necessary to state on whose authority the Regulations are issued and this is usually the Board of Directors or similar body.

Secondly it is important to give definitions of any terms on which there may be doubt.

There should then be a General Section, with what could be called a "clause paramount", on safety, as this should be the first consideration, for example:

"Safety of Ship: The first consideration of every officer shall be the safety of the lives of those on board and of the ship and the cargo. No consideration of programme, convenience, or prior instruction is to be allowed to justify the taking of risk which may endanger the ship or any person on board the ship."

This should be followed by a statement that the Master and officers are to adhere strictly to the regulations of the country of registry of the ship, and the regulations of any country in whose waters the ship is operating.

The Master and officers should also be instructed to give every assistance to port and government officials, and surveyors of the society under which the ship is classified, and any person appointed by the company in relation to the ship.

There should then be a list of the various regulations, codes, etc. placed on board for reference, stating that the master and officers should be guided by them. Some of the codes may not have been adopted or issued by the country of registry, but the shipowner may feel that such are the aims, guidelines and prospective rules of such codes and conventions, that he wishes his sea staff to be guided by them.

There should also be a statement about any departure from the company regulations, stating that the Master should immediately advise the owner if at any time he decides to act contrary to the regulations, giving his reasons for doing so.

Finally, responsibility: It should be clearly stated that Master and officers are responsible for the equipment, safety, efficient navigation, and management of the ship and that nothing contained in the Regulations relieves them from that responsibility.

Considering the two categories of regulations:

Technical: the major problem for anyone compiling company regulations in this category is where to begin and where to stop.

Taking for example the Master and the basic factor of his being responsible for the safety of the ship, crew and cargo: Some shipowners give the Master no instructions at all on this point. If questioned they would probably argue that it is implied, as under the laws of the country of registry of the ship and from which his certificate of competence was issued, he is responsible for the safety of the ship, etc.

Others consider, again probably through bad experience, that it is necessary to state specific areas of responsibility, for example,

That the Master is to:

ensure that the ship has sufficient fuel for the intended voyage before sailing.

ensure that all navigational equipment is in order and that the ship has adequate charts and directions for the intended voyage.

They may also state:

The Master or Chief Engineer may delegate responsibilities to other officers while retaining overall responsibility.

Others go further and stipulate areas for which specific officers are responsible. In some companies this has become necessary through re-allocation of duties resulting from the introduction of new systems of work. That is, where there has been a change from the traditionally accepted duties and responsibilities.

To draw a parallel with the British Royal Navy again:

Under the navigation section of their regulations they make the captain responsible for the safe conduct of the ship, but also say that under his direction the navigating officer is to have charge of the navigation. Thus they emphasise the captain's overall responsibility while at the same time allow him to delegate under his instructions.

Like many of the older British shipping companies, they are specific about the captain's responsibilities approaching, and when in the vicinity of land: they state that he must ensure that the ship's position is ascertained in good time and by the best means possible, and thereafter is constantly fixed.

In the same way in their general instructions to the Staff Engineers, he is to ensure that the operation and maintenance of the machinery and equipment is according to their instructions and to "accepted and professional standards". An interesting term covering something which goes beyond detailed description and can be likened to the "common practice of seamen", which is often used in the Merchant Navy to describe the way professional sailors would do something.

It is not suggested that Merchant shipowners should follow naval procedures in these matters, but the fact that naval authorities consider it necessary to be specific in these matters, despite their highly qualified and disciplined sea staff, is noteworthy.

Until international legislation decrees otherwise, it seems that the decision on whether there should be regulations and how much detail should be in them, will depend on national governments, the attitude of the shipowner, and perhaps the common practice of shipowners of the country of management of the ship, if not the country of registry.

As stated at the beginning of the chapter, many British companies have undoubtedly developed many of their regulations because of court actions associated with interpretation of British maritime law or the possibility of such actions. Perhaps this is a factor which should be considered by those who do not, at present, issue regulations. This is referred to in more detail in Chapter Twelve.

Any shipowner being undecided as to the amount of detail to include in his regulations should not only consider his own protection but the protection and support of those they place in charge of their ships.

Administrative: Some of the items and details in this section will differ from company to company depending upon the concepts of management. These regulations should be specific, not only because the company will want the matters dealt with precisely, but because the sea staff may need to refer to the regulations for guidance on procedures from time to time.

As stated earlier, there are grey areas where an item could be in this section or the technical section, for example: The British Merchant Shipping Acts state that when command is being changed the old Master should hand over all the ship's documents to the new Master. Some shipowners may consider that this is sufficient and say nothing. Others may include a regulation stating that on changing command a list of documents is to be prepared and signed by both Masters. Others may take this one step further and either list all the documents in this section or refer to a company's standard form on which all the documents are listed. Again it is a matter of opinion, but a standard form is helpful to the Master and assures the shipowner that the documents have been properly transferred. Here again experience is a major factor: the owner only needs a ship to be delayed once because a document is missing for such a system to be established.

The following is a list of items, with brief comments, which could be included in this section. It is by no means definitive but highlights subjects on which sea staff look for a ruling from time to time.

Arms and ammunition: whether to be carried or not.

Cost control: purchase authorities and procedures.

Communications: methods, frequency, reporting requirements, urgent and emergency procedures.

Charts and publications: correction and supply system.

Safety and ship management committees: members, frequency of meetings, agenda.

Dangerous drugs: security and protection against misuse.

Direction finder: recording of calibrations and checks on the equipment.

Discipline: The authority of the Master and senior officers.

Fuel safe margin: guidelines on the calculation of fuel requirements for the intended voyage.

Infestation: prevention and control.

Inspection of ships at sea and in port: guidelines on frequency.

International Search and Rescue and Meteorology Organisations, contributions: The company's policy on contributing to the work of such organisations.

Insurance procedures: Accidents to ship and shore personnel, deviation, grounding and similar casualties, collision, damage to property, cargo damage, stevedores damage, hospitalisation and medical treatment of crew members, stowaways, visitors to ships, gangway notices, wives and children in ships.

Cargo claims procedures: Inspection of log books and cargo spaces. Statements to legal and other representatives of the cargo owners.

Hull and machinery claims: procedures to be followed.

Modifications to ship and equipment: authority required before arranging.

Port duty staff: requirements for deck and Engine Officers and ratings.

Night order book: Master to leave written instructions when leaving the bridge at night.

Oil record book: importance or recording details.

Oil transfers: required procedures, precautions and quantities, samples and receipts.

Overtime: guidelines on hours to be worked and controls.

Reports: A summary of all reports required by the company.

Register of derricks and other cargo gear: instructions on the maintenance of records of overhaul and examination.

Salvage: guidelines on the acceptance of assistance in emergency.

Security of information and cash: instructions on procedures to be followed.

Smoking on deck and elsewhere: when and where permitted or prohibited.

Stability: to be calculated for all stages of the voyage.

Stowaways: searches before sailing.

Swimming from ship: to be prohibited.

Testing of equipment before sailing: procedures to be followed.

Vaccination and inoculation: Company requirements prior to sailing for ports where there is risk of infection.

This list can be developed as required for specific ships and trades, but the items above are those on which instruction and guidance may be required on any ship from time to time. No mention has been made of any safety or pollution matters in the list as it will be assumed that a code of practice or similar manual, either published or issued by the company, will be carried. These should be listed in the regulations as described at the beginning of this section.

Motivation in respect of regulations

Despite the issue of well thought and arranged regulations they are, from time to time, ignored and accidents do occur. Such accidents are not confined to Flag of Convenience ships but also occur in ships of major companies with highly qualified officers, safety training programmes, safety and management committees, and detailed regulations concerning every conceivable eventuality.

In such cases what is often lacking is the motivation to follow written or traditional procedures. Without such motivation regulations are as naught. The management involvement in motivating officers and crew will be considered in Chapter Thirteen.

The shipowner's responsibility to ensure that shipmasters and senior officers understand his wishes beyond the written word or regulation will be considered in Chapter Twelve.

Policy

This is now considered in the broader, managerial sense of the word although the definition in the introduction is the same.

Policy in this sense is the statement of intent by senior management which sets the "tone" of the company. Not all companies consider it necessary to state their policy in writing and perhaps would prefer to avoid the lack of room to manoeuvre created by such a statement.

But if a company does have ideals and is prepared to stand by them, then the time taken to prepare such a statement is a worthwhile exercise. Good examples of the type of policies which should be adopted by shipowners can be found in the "Code of Good Management Practice in Safe Ship Operation" issued jointly by the International Chamber of Shipping and the International Shipping Federation. This is reproduced at the end of this book as Appendix One, by kind permission of the International Shipping Federation.
When policies are agreed they should be issued to all employees involved with upholding them. Of course, policies have to have some flexibility, but essentially they should be maintained, and any deviation from them should be approved at a senior level whenever possible. Anyone acting contrary to the company's policy without authority must take full responsiblity for doing so.

The routine policies which are a part of every company's organisation were considered in Chapter Two.

Chapter Five

Communicating, controlling, and informing

"Is there a sound in the forest if a tree crashes down and no one is around to hear it?"

(Ancient riddle, to which the answer is no, there are only soundwaves.)

"It is the recipient who communicates."

The words forming the chapter heading have related meanings in general management terminology and ship management terms as follows:

Communicating

This can be considered in two stages: "communication", the art or form of passing a message from one person or persons to another person or persons and "communications", the way in which such messages are passed.

Communication

There is an old saying that "people hear what they want to hear". This can be expanded to "people hear what they are able to comprehend or perceive". If people do not comprehend what is being said the communication is just a sound wave, like the tree falling in the empty forest. The same applies to a written communication; unless the recipient understands what it is about, it may just as well be in a foreign language. A similar effect can occur if a person is giving too much information which he cannot digest. This will be considered later in the Chapter.

In the early days of the adoption of management techniques, many shipping companies thought that communication between senior and middle management ashore or in ships was largely a matter of keeping staff informed of the company's policies, goals, intentions, and news about the staff, new tonnage, ship sales, etc.

Later, through seminars, shipboard and shore management meetings, voyage conferences, etc., senior management began to "hear" the views of their middle management. The involvement of sea staff in planning and setting their own goals and objectives took this a stage further: senior and middle managers ashore and in ships found they had a number of common perceptions and communication was upwards at least as much as downwards.

Similarly the involvement of petty officers and ratings in the day to day work planning in ships also resulted in better communications "upwards" to the senior staff with improved co-operation, and thus, results.

Of course not all companies were as enlightened as this and there are still those who manage in the old style. But generally, there have been considerable improvements in "communication" in the shipping industry. There seems little doubt that efficiency does improve when all concerned "perceive" the objectives and feel they have a common goal.

The skill of verbal and written communication needs to be developed in most managers, but particularly those involved in ships where a communication "gap" between ship and shore is often exaggerated by the distance between them.

Whereas comparisons with the past are relevant in considering some aspects of ship management, in the field of communication systems one must look to the present and the potential for development in the future.

When already there has been direct spoken communications between men on the Moon and Earth and transmission of computer data from satellites and space craft to Earth, plus the issue of direct equipment operating instructions to equipment in space from Earth, the only limit to marine communications is the time for development and *cost*.

Through INMARSAT and other systems, spoken communications are available on a world-wide basis via the radio telephone, written communication is available via telex and actual copies of weather maps, charter parties, etc. are available through facsimile equipment. Similarly data can be transmitted directly to main frame computers and staff ashore can obtain specified data direct from ships without having to ask for it. Thus there is the capability for the shore office of obtaining the ship's position, as plotted by a satellite navigator, at any time. In the same way operating data can be obtained directly from the engine and other plant and equipment. Television type pictures of areas of the ship, its equipment and cargo can also be relayed directly to the shore office.

Because of the very high development costs, particularly those of the satellites, the initial costs of the equipment and use of satellite time is, as one would expect, expensive. Consequently the shipowner needs to give considerable thought to the equipment he buys, leases, or hires, and the uses he will make of it: as will be considered later in the chapter under "controlling."

As with so much technical equipment, advances are so rapid that the most up to date communications or computer linked data transmission system can be out of date in a very short time. For this reason, apart from the high purchase cost, many owners prefer to hire the equipment.

In time, some of the decisions on the installation of communications will undoubtedly be made for the shipowner through international and national laws. But such requirements are usually minimal in relation to what is available if one is prepared to pay the price.

Thus from a management point of view, immediate communication is possible between ships and the head office and agencies and other organisations around the world.

Data can also be transmitted promptly to computers for storage and retrieval as required. However, it is important to note that if a multi-user computer is used, there will be certain times when it is more expensive than others. But here again technology assists by facilitating automatic communication with the computer at off peak times, if the user is in no hurry. Additionally, high speed transfer techniques minimise costs by transferring data in a fraction of the time taken by telex.

More will be said on the use of computers in the next chapter.

Controlling

Just as there is a need to differentiate between the difference of meanings of "communication" and "communications" so there is a need to be clear about the difference between "controls" and "control" in the management sense.

Control has been described in some detail in *Running Costs* and the analogy of filling a bath and adjusting the taps during filling to obtain a required depth and temperature was used. The measurements of depth and temperature in this case are the "controls". "Control" is the decision making process of how much to adjust the taps to obtain the desired result.

It is most important to bear in mind that controls are usually historical and at best are in the present, whereas control is associated with the future. It is on the information feedback or controls between the ship and other sections of the company, that managers adjust or control activities to achieve planned objectives.

When deciding on the controls required it is important to keep in mind that it is the "quality" of the controls which is important and not the number or frequency. Too many "measurements", too often, are of no value and can confuse as well as being expensive to collect and distribute.

As a general rule controls should be:

Relevant
Timely
Economic
Simple and workable

Considering each of these:

Relevant: Only information which can be put to effective use should be produced. The need for every item should be carefully scrutinised when setting up the system and reviewed from time to time. It is very easy to obtain a vast amount of measurement information. Only the minimum information necessary to control should be obtained, and no more.

Timely: The prompt delivery of information is very important if control is to be exercised in time to be effective. Obviously some controls are more urgent than others and thus the frequency and delivery time of measurements to the manager must be carefully considered.

Economic: Coupled with relevancy and timeliness is the cost of obtaining and delivering information. As has been seen, information can be sent very quickly from ship to shore and vice versa, but when considering how quickly information is required, thought must be given to the cost of obtaining and transmitting it. Mail is relatively quick between many countries, and courier services can be relatively inexpensive, quick, and reliable.

Simple and workable: Like so many aspects of managment, controls should be simple and workable and should take into consideration the equipment and staff available. When setting up controls the information needs of each management section or department must be analysed in relation to their authorities and responsiblities. This should be followed by a flow diagram to check the manner in which the information will move to and fro. (See Diagram 10.)

Controls in ships: Controls in ships are essentially checks against plans which may have been made ashore, in the ship, or jointly, depending upon the management organisation system used as described in Chapter Three.

The prime plan is the financial plan or budget which is constructed upon the following "work" plans related to the company's long and short term plans.

The maintenance schedule or planned maintenance.

The running or operational efficiency "norms", and days in service associated with the operator's requirements.

The manning required for the maintenance and operational requirements.

The overtime associated with the planned maintenence and anticipated operational requirements.

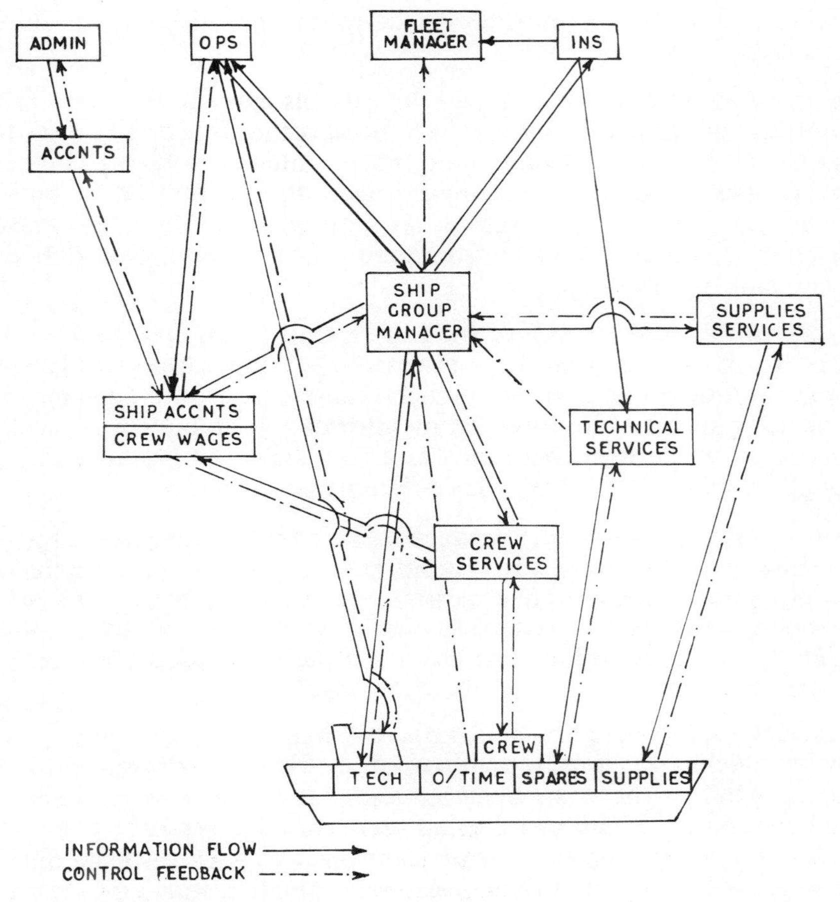

Diagram 10: Information flow chart

Information flow ⟶
Control feedback — · →

NOTE:
The chart is for a centralised, ship group organisation with technical expertise included in the ship group.

The levels and consumption of supplies and spare gear.

The budget controls are produced by putting into the system information about costs as they are incurred. The allocation and arrangement of these costs has been dealt with in *Running Costs*. For the purpose of this chapter it is sufficient to say that information on costs should reach the cost co-ordinating position within reasonable time. As it is usual for costs to be supported by documents this will involve mail or courier service. However, preliminary advice may also be required, and this may be transmitted by telex or computer and put into the accounts as a preliminary entry, to be confirmed or adjusted when the actual accounts are received.

Most of the controls associated with the other plans can be reported immediately within the ship while the speed of transmission of data to the shore base should reflect its importance in terms of taking corrective action. Stocks levels and consumptions need only be provided monthly, or even three monthly, although use of spare gear usually needs to be reported immediately because of the time lag which often occurs between ordering and delivery.

Controls on the performance of the ship, i.e. its speed, fuel and lubricating oil consumptions, revolutions, slip, temperature and gauge readings are associated with the planned maintenance and efficiency. They usually need to be supported by information on outside forces which can affect performance, such as a foul ship's bottom after a prolonged period of inactivity, bad weather, poor fuel, etc.

Methods of reporting on performance such as temperature readings, revolutions, consumptions, vary considerably. Some use direct data access equipment as mentioned earlier, others require daily cable or telex reports. Others are content to await log extracts and other data by mail, before considering whether or not the performance is satisfactory.

Unless very sophisticated machinery is installed with associated diagnostic facilities ashore to help correct inefficient performance, it is debatable whether there is a need for such daily reports at sea under normal conditions. Much will depend upon the organisation of the company, its policy, and the staff in the ships and ashore. If the quality of the sea staff is poor and there are highly qualified staff ashore, then such readings will not only indicate the performance of the ship but also the performance of the staff. However if there is a very small management team ashore and well qualified senior sea staff, then daily reports may not be necessary.

A preferable technique is the one where a range of measurements is given within which no reports are required, but beyond which, a report is required, with if possible, an explanation of the variance. Unfortunately such techniques are rarely popular. To many controllers it is comforting to see a steady stream of information, as it tells them that measurements are being taken and they can see for themselves that all is well, or otherwise.

It is relevant to mention at this point the matter of reporting the ship's noon position: again this varies from daily to weekly, often without any logical basis other than tradition. In some trades such as the banana trade, it is essential that operators know the ships program and ETA so that its arrival can coincide with the availability of cargo, but for many ships there is no such need.

On long ocean passages a weekly report may be sufficient, particularly if the ship is participating in an ocean rescue scheme such as AMVERS and is reporting its position to a central control daily. But essentially it is a matter of choice and the ship owner's peace of mind. If there is no central position co-ordinating organisation, or if he wants to know the ship's progress, or whatever his reasons, then a daily position report can be made. In such case it is usual to include the speed since the last report and any other relevant information. In ships with a satellite navigator associated with automatic data transmitting equipment, the owner can obtain the position whenever it is required.

A less immediate control is also achieved through the records of the ship's activities. The way in which such records are kept and the information contained in them is yet another indication of the management of the ship by the Master and other officers. For this reason the records which are kept should be examined by a supervisor from time to time.

The form of some records may be defined by national or international law such as the Official ship's log, the Radio log, the Record of oily water disposal, the Record of overhaul and Inspection of lifting gear, chains and wires. Others are required by the usual practice of ship management such as the scrap or rough deck and engine logs, and the chief mate's and chief engineer's logs. It is also usual to maintain records of manoeuvring the ship, sounding of tanks, and records of hold or tank temperatures relating to specific cargoes.

The deck log usually records the position of ths ship whenever and however this is obtained, navigational directions and weather conditions, details of leaving and arriving in port, cargo operations, and any matter on which a claim may be made against the ship or others.

The engine log records similar details but with greater numerical detail concerning temperatures, and revolution counter readings and consumptions. Tank soundings are usually kept separately.

Much of the information contained in the scrap deck and engine logs is not of immediate importance. The logs are usually kept in the ship unless required in relation to some legal action or insurance claim, whereas the chief mate's and chief engineer's logs are usually sent to the head office for inspection, and safekeeping at the end of each voyage, or at regular periods. In some companies the scrap logs are combined with the mate's and chief engineer's by using a carbon copy. In such cases the copy is forwarded to the head office monthly or other appropriate period.

Finally there is the control of the performance of staff:

This will relate to all the other controls because if people are not performing properly it will follow that the activities with which they are associated will not be completed in time, or will not be properly executed, or both.

A control can be operated by reports, made by seniors on those who report to them. More often than not these are confidential, although they can be more effective if discussed with the staff member before it is despatched. Whichever way such reports are handled, they should not be filed and forgotten. They should form the basis for appointments and promotions, thus controlling the type of people in positions of responsiblity.

"Control" is really the core of ship managment because unless one is in control one cannot manage. This applies equally to those in ships as well as ashore. The foregoing guidelines should be kept well in mind in any shipping company and should be reviewed from time to time to ensure that they do not grow beyond the needs of the controller.

Although most ship control is exercised through sighting measurement and narrative reports, there is another important control which reflects an old Chinese proverb:

"*the best fertiliser is the farmer's heel*".

No matter how good the system of reporting, there is a need for a manager to see for himself what is going on. This applies to any manager in shipping from the chief engineer in his cabin receiving data log readings, to the manager ashore sitting in his office. From time to time managers need to visit the site of the work and talk to the staff involved. Such visits are complementary to the other controls. They are not quantifiable, but nevertheless give the manager a sense of the work being done and most importantly, an understanding of the staff involved. He will see and hear things which often do not appear in measurements and yet may have a vital influence on the work. In the same way there are certain activities in any company, which although not measureable in the usual sense, need to be checked from time to time to ensure that they are being carried out. This is particularly so in the maintenance of records and the provision of information which will be considered in the next section.

Again the frequency of such visits and the level of the staff making them, will vary from company to company depending upon the organisation system used. In a company with a very small shore staff and ships trading far away from the home base, such visits may be difficult to arrange and expensive. But, they are valuable to all concerned and every effort should be made to see that they take place whenever possible.

Finally, there is the control of resources, i.e. money, staff, and facilities: despite all the planning and associated goals there will be times when managers at all levels have to match their activities to circumstances rather then their plans.

In money terms this is often associated with the cash flow. Although the budget will have been approved and the forecasts to date show the annual expenditure to be within budget, there may be times when the average or quarterly expenditure may be exceeded, affecting the Company's overall cash position. In such cases control has to be exerted to cut back planned expenditures temporarily.

It is probably staff resources which require the greatest control. In many ships today, with small, flexible crews, control is essential. The allocation of staff to the best advantage for the immediate priorities, such as cargo preparation, and the long term objectives, such as the planned maintenance of the ship, needs careful consideration. In this, decisions taken jointly by a management committee usually produce the best solution.

Ashore, allocation of staff resources also needs control, e.g. superintendents visiting ships, accountants working on end of year accounts, holiday requirements, etc. Despite all the time spent in defining the workload of the company and the staff required to deal with it, there will always be fluctuations which require control decisions from managers on the allocation of staff. (See also Chapter Two.)

Few companies can afford to have enough staff for the peaks of activity and usually arrange for average activity. As will be seen later in Chapter Fourteen, one of the problems of ship manning scales is the need to man for peak activities and then finding ways to utilise staff effectively at other times.

Again, although equipment such as photocopiers, word processors, etc., may be adequate for the company's overall needs, the demands upon it may be irregular such that decisions will need to be taken as to priority of usage. An example of this is access to the main frame computer.

The controllers: Whereas every manager controls to some degree, the place and position of staff who make control decisions depends, to a large degree, upon the system of management, i.e. centralised or decentralised and the involvement of staff in the plans. If the plans are made wholly by the shorebase then they must control, although they may delegate some of the control to the ship. This does work, but is not so effective as when the sea staff are involved and take responsibility for the plans and their execution. As explained in Chapter Three, in such cases the role of the shore staff has a different aspect, being essentially supportive and advisory, although they too have planned activities which need to be controlled.

Informing

As has been seen, controls are information about the performance of ships, equipment, and staff. But there are other forms of information which pass between ships and the shore which can be termed "Operational", "Managerial", and "Advisory".

Operational: Essentially this is information which is passed between ship and operator or agency and vice versa, which will usually include the following information:

From the ship
The estimated time of arrival at the next port, usually given at specified intervals.
Bunkers remaining on board on arrival at ports.
Estimated time of departure from ports.
Details of cargo operations.
Reasons for delays.

From the shore
Berth and cargo arrangements prior to and on arrival.
Voyage and cargo carriage instructions.
Bunkers arrangements.
Agency details.

Managerial: This is usually information between ship and shore relating to crew movements and arrangements for supplies, spare gear requirements and arrangements regarding repairs. In companies where the ship's staff are involved in or accountable for the costs there will also be a flow of information back and forth on such matters.

Advisory: This type of information usually passes from the functional departments ashore to the ships, although ships themselves may sometimes be a source of information, such as when reporting navigational hazards or experiences with difficult cargoes. The type of information usually issued is as follows:

Crew information relates to changes in the conditions of service, e.g. new wage and salary structures, leave rates, etc. But it may also include advice on crew health, repatriation regulations, and immigration and immunisation requirements of different countries.

Navigational information is issued by the government of the country of registry or the government of one of the maritime nations and gives warnings of hazards, changes in navigation warnings, etc. The shore office acts mainly as a distribution centre through the Technical Department. (See also Chapter Nine.)

Port and seaway information is issued by port and other authorities and again distributed from the head office. This information covers such matters as drafts, requirements as to manifests, crew lists, etc., crane and repair facilities, availability of tugs, bunkers, and fresh water. Some publishers co-ordinate and publish complete books of such information, supplementing the existing Pilot books.

Supplies information concerns food prices and commodity availability, agreements with suppliers, victualling scales, relevant local regulations, certification of stores, etc. (See also Chapter Ten.)

Since the 1960s safety has been a matter of particular concern. Information on this subject is usually issued by government departments or industrial institutions and distributed by the shipowner. The notices and advice can cover a wide range of subjects and include examples of specific accidents of which seafarers should take note. Warnings of possible dangers in using some equipment and commodities and in the carriage of specific cargoes, are often contained in notices issued by Governments, but may also be issued by institutions or the P & I Clubs.

As one would expect at times of considerable technological development in an industry, there is also a considerable amount advice issued on equipment and methods of operation. In this area more than any other, the manager has a difficult task in deciding whether to select information relevant to the Company's ships and equipment only, or whether to provide more information through magazines, etc., to stimulate a wider interest.

Supplies information usually concerns prices and availability of stores abroad, but can include special storing policies and advice on contracts with specific suppliers on a world wide basis.

Insurance information is usually of a legal nature, concerning such matters as procedures with Bills of Ladings, accidents to crew, people visiting ship, etc. Procedures to avoid claims will probably be issed by the relevant functional department on the advice of the insurance department. (See also Chapter Eleven.)

In all these areas care needs to be taken in controlling the information being passed to ships. Not only has there to be a decision on what to pass, but also a system to ensure that ships receive the information and that it is kept up to date. The manager must consider that he has a responsibility to keep ships advised of anything appertaining to them. In this he must make sure that he is also "informed". He should consider the effects of sending any information to ships, i.e. what the ship's staff will do with it when they receive it, and whether he should add any comments to the information, or just send it with a complimentary slip. This is also referred to in Chapter Four.

Chapter Six

Collecting and storing information

Records: The reasons for information from ships have already been considered in Chapter Five, i.e. it is required for both legal and control purposes. The question of how to collect and keep it requires careful thought.

Today, information is collected physically, and automatically. Physically by staff taking or reading measurements and noting actions and events. Automatically through data loggers recording temperature, revolutions, course changes, engine speed alterations, etc.

The way this information is transferred for control evaluation and record can vary considerably, and will probably continue to do so for many years, because of the large difference in costs of the systems involved.

At the one extreme the information is recorded on forms and some of those forms sent by mail or hand to the head office as required. At the other extreme a data logger in the ship is interphased with a shore computer and other sophisticated transmitting equipment such that the information is automatically collected "untouched by human hand".

Between these two extremes are a number of alternatives:

Selected information from forms sent ashore is stored in a computer in the head office. Selected data log sheets are mailed to the head office.

Selected data is stored in a mini-computer in the ship either automatically, or by hand. "Copies" of the entries are then produced on floppy discs or diskettes and mailed to the head office.

As the cost of automatically recording and transmitting information from ships is likely to remain high, and the resultant data format does not suit every purpose, forms will undoubtedly be with the shipping industry for some years to come. In any case, lessons learned about forms are applicable to other forms of data collection and recording, and for those reasons forms will be considered first.

Forms

The first thing any manager has to decide is what he has to record for his protection against claims and what he needs to know to fit the system of control.

He next needs to decide the format of the records and reports. There is an old organisation and methods (O & M) maxim which says that any form, or report, should be as much use to the person completing it as the person who gets it. In other words the person completing the report should be able to use it too. He should not have to produce the data in one way for the controller and produce it in another way for his own use. The form should be of working use to them both.

Ideally any form should be multipurpose, i.e. the same form should be capable of being used by a number of different departments even though they may not all require all the information recorded. If for some reason this cannot be, then at least the order of recording the information should be consistent. It is very easy to make mistakes when transposing columns of data from one record to another if the columns are in different order.

Forms also need to be controlled. No matter how well they are thought out, the need for them and their contents tend to change with time. For that reason all forms should be reviewed on a regular basis to ensure that their contents are still required. Any company using more than a few forms between ship and shore needs to make an inventory of all the forms it uses. It should establish who completes them and who uses the information contained in them, to ensure that the information is really needed and that there is no duplication of usage and effort.

Much of the recording of events and measurements in the shipping industry is of a standard nature, as described in Chapter Five. For this reason a number of publishers and consultancy firms produce standard deck and engine log books, and planned maintenance and spare gear control systems, which can be adapted to individual ship needs.

Thus, shipping companies do not have to involve thenselves in time consuming and expensive research into the basic information they require and design of the associated forms. They can, if they wish, adopt an "off the shelf" system which in many cases may be adequate for their needs and will fulfil any "legal" requirements of record keeping. Similarly there are standard computer software packages which can provide a reasonable ship control system without months of design time for an owner's particular needs.

For the large sophisticated shipping company which seeks greater efficiency through tight control, the design of special forms and computer programs to suit their needs can have advantages. But such companies need to guard against the unnecessary growth of information requirements and always seek a standard format in preference to a specifically and costly designed format. They should also guard against those who are more interested in the systems than in the savings or efficiency they should bring.

Computers

As was stated earlier in the chapter, many of the lessons on forms are applicable to computers and forms may still be required to collect data to put into the more simple computers. But beyond that, the computer does have considerable advantage over other systems of collecting and presenting control data. It can:

Store large amounts of information in a very small space.

Provide access to common information for staff thousands of miles apart.

Allow input of information by staff thousands of miles apart.

Analyse and present data in various different ways.

Maintain inventories of supplies and equipment, automatically up-dating the record when supplies, etc. are used and automatically preparing replacement orders.

Produce information on a screen for review, or print out information as required by outside organisations, such as auditors and industrial federations.

Provide security checks such that competitors cannot access the information and even arrange that access to certain records, etc. is restricted to designated personnel.

Although small computers are now relatively cheap and readily available, like all communicating, control and recording systems *they demand that the users know what they want*. If the users do not know what they want, then the computer can become just as complex and superfluous as the conglomeration of forms which were often found in many of the older shipping companies in the 1950/60s.

Files

Another old O & M maxim is that most of the things one wants to refer to are in the top half inch or centimeter of the file. This is undoubtedly true, although in shipping one never knows when one is going to have to look much further back in relation to claims and associated events. Thus care is required in filing.

Essentially the contents of files fall into two categories: information and documents. The information is that which is usually recorded on forms or fed directly into a computer, or a combination of both. The documents can range from survey reports, certificates, letters and cargo plans, to repair and supplies accounts. The documents themselves need to be retained intact for legal purposes but much of the information can be transferred to a computer for ease of access.

There are no particular rules for filing documents in a shipping company, except for those of logic and the need to find documents and information as quickly as possible. But no matter how carefully a filing system is created there will always be the factor of the human element to contend with, where different people will put the same letter into different files because they see its content in a different way. If in doubt a copy should be put in each. Many shipping companies try to avoid this problem by insisting that each communication should only be written on one subject.

At the head office, files will usually be kept by department, and ship names. In ships the filing system should, preferably, match that of the shore organisation. In a company with regular senior officers transferring from ship to ship it can be an advantage if there is a standard system in all the ships.

There are two final points on this subject:

Numbering: Although not always essential, the consecutive numbering of letters, forms, etc. can be of assistance in shipping where letters and other documents are more likely to go astray than in other industries.

Disposal: Most people in business can tell you how to file but it is rare to get any instruction on when to dispose of documents in a file. All documents associated with accounts, including contracts, charters, specifications, must be available for most company auditors before finalising the annual accounts. Thereafter there is usually a Government Tax rule requiring their availability for at least three years by most countries.

Other documents relating to the ship's technical history and voyages should probably be kept much longer, as these may be important evidence in refuting claims. In many cases there are no specific rules but as a broad guideline, it would be unwise to dispose of documents less than five years old. This should not prevent removal from working files to a second stage filing or storage, where they can be bound together or boxed, providing there is a system for their retrieval if required. In certain cases micro-film copies can allow immediate access of data from records, etc. while the actual files are stored elsewhere.

Caring for documents and information in an important function of the shipping company. Neglect of this care, can not only increase the work load through wasted effort in searching for information or documents, but can be very costly if important documents are lost.

Part Two

CONCLUSION

Shipping is an information industry. There is so much information issued and required in various forms that management has a responsibility to see that it is all relevant and understood by those who obtain, receive, and use it.

Regulations should be complementary to existing government laws and industrial practices, providing staff with further guidance in safeguarding the ship, life, property, and the environment. They should provide guidance for sea staff on procedural and other matters involving the protection of the shipowner's interests.

Policies set the tone of the company's management and although often broad in description should be sufficiently clear for staff to follow.

Communication depends upon perception and the ability to hear what is being said. Management should listen to what their staff have to impart, thereby obtaining a broader knowledge of the work and satisfying the need of staff for involvement.

As was seen in Chapter One, the development of communications was a cause of early changes in the industry. There is now a capacity for almost instantaneous and unlimited transmission of information.

Controls are at the core of ship management. But the "need to know" what is happening, the frequency of sending controls information and the uses to which it is put, needs careful control in itself. Gathering, transmitting, and dealing with information can be expensive, not only if it is transmitted by high cost technology, but also in terms of the man hours involved.

Despite all the technical assistance in controlling, the manager should not neglect to go and see for himself what is going on from time to time.

The care of information and documents is a very important activity of the industry. The lack of care can be costly, not only in the time spent in searching for information and documents, but also if they cannot be found.

Motivation in adhering to regulations and policy and assisting in control activities is vital and is helped if staff are involved in their formation and clarity.

Chapter Seven

Administration and accounts

Administration

Whereas the role of the ship management shore staff is to support the ship, administration supports both the whole shore staff and the ships.

The word administration has two meanings in business: that of senior management, i.e. the administrators and administration, or "admin", the provision of office services.

The role of senior management has been described in Chapter Three. The purpose of this chapter is to consider the service type of "administration" which for budget and accounting purposes are often described as "overheads". Col. Urwick once wrote that there are no overheads, only contributions and this correctly places the position of administration in any business. For in any business there are a number of basic tasks concerned with just running an office and unless someone is appointed specifically to do them the work will have to be done by the specialists in the line and functional departments, thus reducing their effectiveness.

The size of the job and the calibre of the person required to manage the administration will usually depend upon the size of the company. In a very small company the owner will probably delegate much of the work to his secretary, in a larger company the chief accountant or company secretary may take on the task, or an "office" manager will be appointed. In large organisations there will be an administration department and perhaps a separate personnel department.

It is not the purpose of this chapter to dwell on this function in any depth, as it is a function of all businesses and there is no special part which only applies to shipping. The point of introducing the subject is to highlight the need for someone to carry out the function in any shipping company. The work associated with it can be very time consuming and thus it cannot be ignored.

Briefly the scope of administration is as follows:

Premises:
Purchase or lease contracts
Heating, cooling, lighting
Cleaning
Maintenance and decoration
Fire protection and safety
Repairs and maintenance.

Furniture and equipment:
Purchase or lease contracts
Repairs, maintenance and renewals
Telephones, telex, typewriters, photocopiers etc.

Shore staff:
Engagement and dismissal
Conditions of service
Holiday arrangements
Sickness
Temporary staff
Medical and sick leave.

Training:
Pensions and insurances
Luncheon varieties
Beverage facilities
Company cars.

Supplies:
Stationery
Cleaning materials
Beverage.

Insurance:
Staff and office protection
Accident record
First aid facilities.

Miscellaneous:
Mail distribution and dispatch
Periodicals and newspapers
Hotel and travel arrangements
Ticket and visa collection
Messenger services
Car hire and taxis.

The most direct link of the department with ships is usually the very important role of dispatching and forwarding mail.

Accounts

Like administration, the accounts function can be found in any business and the size of the department will usually reflect the size of the company.

Again the role is supportive, although there is an advisory side to it. As mentioned in *Running Costs*, there are a number of accountancy practices which need to be followed regardless of the way in which other managers see the accounts process.

The computer has facilitated the work of the shipping accountant and allowed speedy production of data in a variety of forms, plus easy access to that data by both accountants and managers. This is most important in any organisation where staff are closely involved in costs, particularly those practising accountability to the full.

From the point of view of the ship manager the Accounts Department's role is that of a co-ordinator and distributor. Receiving approved budgets and forecasts, actual costs and estimates and processing them, such that the financial information requirements of all levels and departments of the organisation are satisfied.

The role of the accountants is usually both professional and legal and they are restrained accordingly as described in Chapter One. For this reason at least the head of the department should be a qualified professional, although he may be supported by assistants and bookkeepers without qualifications. In a very small shipping company a part-time qualified accountant or a firm of accountants may be used instead of employing a full-time professional.

Of particular importance in a shipping company is the task of dealing with crew wages and salaries if the shipmaster or recruitment agent does not handle them.

Ensuring that money is promptly and properly transferred abroad is another activity of the accounts department with particular significance in shipping, where ships can be delayed through lack of funds. Crew unrest can also be caused through non-payment of wages.

Chapter Eight

Crew

There are four ways to arrange crews for ships:

Direct employment of personnel by the shipping company.
Employment of personnel through a national organisation.
Employment of personnel through a union.
Employment of crew through an agency
or, a combination of any of these.

Some companies have little choice of the way in which they recruit, engage, or employ their staff, either because of the laws of the country of registry, or union or national industrial influence, i.e. a federation.

The organisation and work of the Crew Department is influenced more by these factors than by the type of management organisation ashore although in some companies with simulated decentralised management systems, the ship's staff are closely involved in where and when staff are changed, and in some cases, the numbers on board. Despite this the actual arrangements must, of necessity, be made by the shore staff.

In the case of a shipping company employing all its staff directly the full personnel function is required and this is now considered in the order in which the work usually occurs.

DIRECTLY EMPLOYED SEA STAFF

It has been said that personnel work is very akin to hospital work, i.e., there is a considerable amount of paperwork but there is also the work of caring for people. Although seafarers spend much of their working time at sea and are, therefore, under the care of the shipmaster, when ashore they are closely associated with the crew department. Similarly when at sea, their next of kin will usually approach the crew department if they have any problems such as family illness, non receipt of money, etc.

The amount of care will depend, to a large degree on the company's policy and can range from the impersonal to the indulgent. Unfortunately such approaches sometimes reflect the state of the industry, i.e., whether there is a high employment or unemployment. In fairness, today many enlightened shipping companies do their best to care for their regular staff, even when forced to dispense with them as ships are sold. They do this by making the best possible retirement or redudancy payments and assist them in seeking other work.

Considering the principal activities of the crew departments:

The manning scale

The first thing which has to be established is the numbers and categories and qualifications of seafarers required to sail in each ship. There may well be minimum requirements imposed by the government of the country of registry, the industrial federation or unions, and these will have to be borne well in mind. Similarly there may be technical or operational requirements necessitating additions to the basic scale. In some cases reductions in the scales can be achieved through labour saving and safety systems and associated agreement with government and, or, union approval. Whatever the criteria, the scale must be established and approved by those managing the ship. In the case of a centralised management, this would be the technical, supplies and operational staff ashore. With a simulated decentralised system the sea staff should be involved in the decision.

The next step is to decide on the establishment:

The establishment: Is the total number of sea staff required to crew all the ships in the company's fleet, and will include seafarers at sea and ashore for any purpose. It is usually based on the manning scales of all the ships plus allowances for leave, sickness, study for certificates, training, and overlap when changing crews. These will be in accordance with the conditions of service of the company and may vary from rank to rank or rating depending upon the leave and study requirements and even the likelihood of sickness amongst senior staff.

Agreeing the numbers to be employed is the first step and may well involve approval of senior management. Some companies lead a hand to mouth existence in this, deliberately keeping well below their known establishment requirements, while others endeavour to keep as closely as possible to a theoretical requirement. For more detail on this subject see *Running Costs*.

Recruitment

Once the establishment has been decided the next step is to obtain the staff and maintain them. The amount of effort and expertise required for this will vary from country to country and from time to time, reflecting employment levels and standards of living, i.e., the higher the standard ashore, the less likely are people to be attracted to a life at sea. There are, of course, always some who enjoy the life at sea for its own sake.

But regardless of this, seafaring does involve a certain amount of deprivation which although helped by allowing long leave, wives on board ships, and other amenities, does not wholly compensate for the apparently normal lives of friends and relatives ashore. Thus there is a constant attraction away from the sea, particularly when staff become married. This attraction is greater when there are many job opportunities ashore and vice versa.

The need to recruit seafarers in general varies, but is always greater for those with qualifications and specific skills, such as electrical and refrigeration engineers, who can be readily used elsewhere.

In some cases a market place type of recruitment is operated; the work goes out that the company needs a particular type of seafarer and eventually a number of such people apply for the job. Others use centralised employment organisations where both employer and employees can be brought together.

One of the most common methods of recruiting directly employed staff is by advertising in local and national newspapers or in professional and industrial magazines, journals, or newspapers. But advertising can be an expensive business and care needs to be taken that the advertisements are placed where they will have the greatest effect.

Having obtained applicants for the jobs the next stage is the interview:

The interview: In this, shipping is no different to any other industry. The level of the interviewer in the organisation, the sophistication of the interview techniques, the background and referee requirements of the company will vary considerably.

Anyone being considered for direct, long term employment should be interviewed. Similarly the more senior or important the appointment, the more stringent should be the interview and checking process, even if the appointment is temporary. The persons being chosen for key positions are to be placed in charge of lives and expensive equipment and the interview process should reflect the company's care in making the choice. (See also Chapter Thirteen.)

One of the great difficulties of the shipping industry in regard to interviewing is the "ship must sail" situation, which is probably unique to the industry. In this situation there is a shortage because, say, a Chief Engineer has been taken ill abroad and put ashore into hospital. The ship is due to sail in two days time but cannot sail without a certificated Chief Engineer. The second or first assistant engineer does not have a chief's qualifications, and there are no chiefs available ashore on the company's establishment.

The work is put out through various channels in the industry for a replacement but only one applicant appears. He is properly qualified but his career has been spent changing from company to company.

The crew department are unhappy about the man, perhaps through a hunch feeling and so perhaps is the technical department. Nothing can be proved, but under normal circumstances time-consuming checks would be made before the man is engaged. In this case he is engaged because of the potential delay to the ship. This is a situation which has to be faced by any crew department from time to time, and can only be answered by considering what their position, and that of the owners, would be in the event of an accident as a result of the appointment.

Although the interviewing of the key sea staff is a responsibility of the crew department, it is preferable that they should be joined by an appropriate member of the technical or supplies department for the interview, so that the applicant's professional qualifications and experience can be considered.

The interview is, of course, a two-way process and the applicant will want to know about the company, his or her career prospects, and what is being offered to join it. This should be provided via a copy of the company's "Conditions of Service":

Conditions of service

The way these are produced can vary from a printed, glossy booklet, to a typewritten set of papers. Regardless of presentation it is the content which matters. Apart from any legal requirements of a country which may require a written contract, the seafarer should be clear about what is being offered. It is a responsibility of the crew department to propose conditions of service and amendments to senior management, and if necessary, seek union approval for them. Thereafter the department should aim to present the conditions in a clear and concise form.

Essentially the Conditions of Service should contain:

Salary scales: For all sea staff although a separate set of conditions may be produced for officers and ratings. In some cases, such as when recruiting through a federation, the federation may produce the conditions of service.

In relation to salaries, the scales should show the "begin at" rate and also any seniority service increments. They should also state how such increments are to be applied, i.e., whether seniority is to count from the time a person first held the position in any ship, or whether the seniority service time only counts with the company. Similarly, whether service includes leave time or not.

Leave scales: The number of days leave earned for sea service and whether leave pay is included in the pay for service in the ship, or paid separately at the end of the voyage.

Other allowances: These will vary from country to country and often reflect the age of the industry in a particular country. They will probably include such items as subsistence allowance when travelling to and from ships and on leave, study leave while studying for government certificates of competency, service bonuses and payments for additional qualifications.

Benefits: Will include details of insurance and pension arrangements made on behalf of the seafarer and state whether they have to contribute towards them or not.

The engagement

Both seafarer and owner having agreed to work together, the next responsibility of the crew department is to issue a letter confirming the engagement. When this has been acknowledged by the new employee, notification should be made to the wages department if the wages are to be paid by salary, or to the shipmaster if he is responsible for the payment of wages. Details of these different systems of payment are included in *Running Costs.*

The prompt payment of wages and other cash items such as travelling expenses, is a vital part of any shipowner-seafarer relationship and the seafarer needs to feel comfortable that payments to his family or bank are being made properly. It is therefore very important that there is a smooth flow of information regarding the seafarer's status, between the crew department and the wages department. Depending upon the system of payment, i.e., salary or voyage pay, the wages department will need to know of the date of commencement of pay and any subsequent changes in rates, and other expenses resulting from:

Annual or other increases or decreases in salary scales.
Individual seniority increments.
Promotions.
Standing-by awaiting appointments.
Sickness.
Training for company requirements.
Leave.
Subsistence on leave or elsewhere.
Study pay for statutory qualifications.
Increments on gaining additional qualifications.
Inclusion in any pension scheme.
Approved travelling and other expenses.

The appointment

Having engaged the seafarer the next step is to appoint him to a ship.

The greatest responsibility of any crew department is keeping ships manned. Before the advent of the computer one of the most effective ways of arranging staff was through the visual appointments board, where movable cards or plaques were hooked onto position spaces for each ship. Similar position spaces and hooks were arranged for staff on leave, training, sickness, etc. Thus the crew department staff could see at a glance the status of the manning of each ship and the staff ashore, and could decide when and where changes could be made.

The computer has accelerated this process and provides an additional aid with more information than feasible on an appointment board. It can still "simulate" the board on a screen but can also select those available by categories of age, current leave earned and taken to date, experience in the type of ship, qualfications and training etc.

Crew changes

Another major task of the crew department is arranging for crew members to join ships, to replace those due for leave or leaving for other reasons.

Any crew changes must be economic and practical. Changes should be made with the maximum number of sea staff at the same time to obtain beneficial discounts of travel costs and reduce repetition of effort in making such arrangements. They should also be made as close as possible to the place from which crew are engaged to minimise costs.

Unfortunately the ship's trading pattern may not always allow changes near to the place of engagement, and for technical reasons it may not always be advisable to change everyone, or large numbers, at the same time. This may well happen in the case of senior or key officers, when to take all the men from one department at the same time may leave the ship without staff experienced in its particular characteristics.

For these reasons, close co-ordination is required with the operators on when to make changes and also required with the technical staff on how to make the changes. Depending upon the size of the company, a system of seeking technical approval for appointments and changes should be arranged. This applies particularly when promotions are proposed.

Invariably there will be some "overlap" in the changes, i.e., time when two crew members virtually fill the same job or are charged against the ship's account when travelling and actually taking over from one another. Except when required for technical reasons, such as familiarisation with new equipment, the crew department should endeavour to minimise this overlap time as much as possible.

Expenses

Unlike shore workers, most seafarers are usually paid expenses to join and leave their ships, unless the travel arrangements are made for them.

Any such personal expense should be approved by the crew department before payment is made.

Expenses need careful control and the amounts or allowances should be laid down in the conditions of service or elsewhere, so that "interpretation" can be minimised. This should be a policy matter and will probably be associated with industrial or union agreements.

Study leave and training

It is important at this point to make the distinction between study leave and training.

Study leave: Is leave allowed for the study for mariners' certificates of competency required by law in each country. The student usually requires to attend at a marine college or academy, even though he may have studied at sea through correspondence courses or other means. In some countries attendance at the college is obligatory and in many countries there are agreed "scales" of study leave for each certificate.

Countries vary considerably as to who pays for this time of "non-earning"; the seafarer, the owner, or both; i.e. the seafarer takes a reduction in pay, but is supported by the shipowner. An additional factor is the relationship between the time served with a particular owner to the amount of paid study leave the owner is obliged to provide. In many countries there are rules to cover cases where a seafarer goes from company to company but nevertheless wishes to study for his qualifications.

Applications for study leave need careful attention. Apart from the fact that the owner may be compelled to allow the seafarer to take study leave and pay him for the duration, he may have a vested interest in encouraging his long term employees to progress their careers. On the other hand it is usually difficult to plan study leave. It is often the officer's own decisions as to when he takes the leave and this is not always compatible with other crew change plans. There are thus, often, conflicting interests within the department itself.

Training: This can be split into two groups, "commencement" and "on going".

Commencement training: Today few seafarers join ships without any pre-sea training, such as the old system of deck boy or cabin boy straight from school. In many countries the rating's pre-sea training is provided by the industry as a whole, or the government, and a levy or charge imposed on individual owners in relation to the number and type of seafarers they employ. The ship's staff and crew department are often involved in such schemes because of the need to follow up the initial training with progressive training in service. This usually requires reports showing time spent exercising particular skills, such as in steering and maintenance of specific items of equipment and levels of competency.

The training of future officers through cadet or apprentice schemes is much more complex and often reflects the development of the maritime industry in a country, and whether the training is directed towards the future supply of officers for a particular company or the industry as a whole.

In many of the older maritime nations, the days of taking a young person straight from school, appointing him to a number of ships until he had acquired the requisite sea time to sit for his examinations, and then leaving him without pay for the period of study for this first certificate are long gone.

There are many cadet schemes, often involving "sandwich" training with periods in college or workshops and at sea before the first appointment as a ship's officer. In some countries the theoretical part of the examination is passed before the future officer goes to sea, although the final certificate is now awarded until "sea experience" has been gained.

The role of the crew department in this depends directly upon whether the future officers are "company" or "industry" trained. With a company scheme the work is considerable and in large companies necessitates the establishment of a separate cadet or training department.

If the cadet is "industry" trained the crew department and sea staff are still involved to some degree because of the need to monitor the cadet's progress in the ships. However, they are at least relieved of the task of the arrangements for attendance at technical colleges, finding accommodation for them and liaising with parents.

But whether the cadet is trained by a company or the industry, careful selection is vital, and should preferably involve aptitude tests, if losses resulting from their leaving during, or shortly after training are to be avoided.

Once the cadet is selected, he or she also needs to be carefully monitored throughout the training, because of the tendency of young persons to be uncertain of their career choice, requiring support and counselling by an experienced person.

On going training: This also reflects the changes in the industry, again varying from country to country. In many cases, courses which were at first supplementary to statutory requirements and thus voluntary, such as those associated with radar operation, and fire fighting have become obligatory and form part of the certificates.

Others, associated with specialised equipment such as turbo blowers and new concepts of management in ships, remain voluntary. With these the shipowner has to decide on the courses which will assist his sea staff in the operation of his ships. Before he can make the necessary policy decision, he has to be fully briefed by the crew department on what the courses will provide and their cost, with assessment of their value from the technical staff, and advice from the sea staff to whom such courses are most appropriate.

A policy decision on this training is necessary because the courses are not cheap. They usually involve the costs of travel to and from the place where the course is held, the cost of accommodation, the course fees and the cost of taking the officer out of service. There is also the cost of the waste if the officer leaves the owners employ shortly after receiving the training. Thus individual consideration needs to be given to those who are sent on courses.

It is noteworthy that alternatives to taking sea staff out of ships for training ashore can be provided by video films shown in ships, supplemented by on board discussion groups. Additionally some companies place specialists in ships to train crews in safety and anti-pollution techniques.

Thus the crew department have an involvement in training and need to be conversant with facilities available together with costs, course content, etc. Once a policy on training has been agreed it is for them to implement the policy within its parameters e.g. that such and such a course is essential or desirable for second engineers sailing in ships with a particular turbo blower. Or that a management course provided by a particular college is appropriate for all senior sea staff sailing in ships with a general purpose crew and a shipboard management system.

Selection of officers for such courses is aided by a computer and can be associated with the appointment of officers to ships.

Employers federations and unions

The employers federation enables collective viewpoints to be presented to governments, unions, and other bodies. The unions in representing their members give their views to governments and employers.

In a country with a large shipping industry, membership of an employers federation, if it exists, is usually necessary if the shipowner wants to participate in such collective representation.

As far as crews are concerned the work of an employers federation can cover a wide field. Not only will it involve negotiations with unions at a national level in regard to wages and other conditions of service, but may well involve training, manpower planning, recruitment and the supply of certain categories of seafarers. In some of these it may represent the shipowner on a Joint Maritime Board including unions in its membership, where policy matters of mutual interest are decided. It may also represent the shipowner in discussions with government bodies regarding qualifications, industrial education, subsidies for shipping and tax matters.

Again, the federation may represent the owner in other employer/union bodies which have been created to deal with pensions and other social benefits for seafarers.

Thus by belonging to a federation the shipowner has collective representation in all these matters. If he does not like the collective views he has an opportunity to put his point of view forward in committee. But it is the majority viewpoint which will be expressed.

The owner does not have to involve himself directly in the activities of the federation, but as a member he will be expected to provide representation on a number of committees. A member of the crew department may be delegated to join some of these committees or at least brief a senior member of the company attending them.

As to the unions, the owner's involvement may well be indirect and direct. The indirect relationship will be in regard to wages and other conditions of service negotiated at national level through the federation.

Directly, the owner may deal with the unions on all matters regardless of whether he belongs to a federation or not. In this both he and the unions may argue that his ships are a special case and the nationally negotiated pay scales and leave agreements are only used as a base upon which to construct a special "package". Or he may only deal with the unions in small disputes over such matters as dismissals or overtime claims.

There can be a number of variations to these themes and the work of the crew department will vary directly with the owners policy for direct involvement with the unions. Assuming that an owner wishes to negotiate directly with all the unions concerned on all conditions of service, including redundancy payments and special agreements associated with reduced manning, then the work load of the department can be very high at times of negotiation.

The actual negotiation, probably headed by a board member, will only take a relatively small amount of time. The bulk of the work will be in preparing and costing the proposals and then costing the implications of any counter-proposals and seeking a compromise solution, again in terms of costs as well as feasibility.

When shipboard systems are to be changed, such as with General Purpose or Interdepartmental Flexibility manning, there will be a number of stages of involvement with the unions. Firstly the proposal and terms of reference for the current work on ship to be studied, then the proposals resulting from the study and the remuneration for the staff for any "change", and finally the implementation of the new system.

Usually crew department staff and union officials work fairly harmoniously together. Each has his job to do and very often they are fully aware of and sympathetic to the others problems. More often than not the crew department have difficulty in trying to persuade senior management to accept new proposals which they consider will benefit the company, while the union officials have their problems with extremists.

Inquiries, warnings, and disciplinary action

The gathering of facts is an important factor in this. If professional competence is in question, the crew manager should be joined by the technical or supplies manager in inquiring, warning, or making recommendations.

If the matter is also being considered by a government department or in a legal action, care needs to be taken not to pre-empt those procedures.

Whenever possible, it is always preferable to counsel staff whose performance is unsatisfactory, before the need for disciplinary action arises. This often prevents the need for further action.

Dismissals

In many countries care needs to be taken in dismissing any member of the staff if the company's position is to be protected in any subsequent legal action. It is therefore important that the crew department should be fully aware of the legal requirements for dismissal in the country where the contract originated.

In the case of misbehaviour in ships, it is important that the crew departments brief the senior sea staff on procedures to protect the owner. Then, if it is decided that a seafarer is to be dismissed, the crew department can take the appropriate action. If for example, warnings have been given to someone who has repeatedly misbehaved and these have been properly witnessed and recorded, the department will be helped in their task.

Crew employment laws

In addition to the laws on dismissal, the crew department should be conversant with the legal requirements associated with the country of origin of the crews they employ. These may range from the social security payments, sickness benefits and medical and dental treatment, to repatriation and pension schemes, and even death of a seafarer abroad.

Pension and insurance schemes

In addition to state and industrial pension schemes, some shipowners provide pension or annuity schemes for their regular employees, with or without a contribution from the employee. They may also provide free insurances for death in service, widows pensions and long term sickness benefits. Again the crew department need to be conversant with all the rules of such schemes, particularly as they affect staff joining and leaving the company. If the company has its own pension fund a member of the department may be required to sit on the board of the Fund.

Welfare

Finally, reverting to the hospital analogy, there is the work of caring for the sea staff. This does not mean that they have to be nursed, but they do need to be able to talk to someone from time to time about their career, or any personal problems they have affecting their work.

Listening and counselling is a time-consuming business but when employees are out of touch with their base for months at a time it is an essential requirement of any crew department.

The welfare role also involves dealing with seafarers or their families at times of serious illness, accident or death. In this the role of the crew department is necessary but difficult.

Advisory

As has been seen, it is a function of the crew department to advise senior management on all personnel matters and particularly to provide information during union negotiations. Similarly they should propose policies on the conditions of service of the sea staff and working arrangements in ships in conjuction with the technical department.

But they must also keep the sea staff informed on matters affecting them, such as changes in the conditions of service, industrial agreements and legislation. This advice is particularly necessary for the shipmaster who takes on much of the personnel function when the ship is away from the home port. (See also Chapter Five.)

Most importantly they must keep the accounts department promptly advised of any change in the status of the sea staff, in order that adjustments to salaries and other payments or deductions be made.

Such then are the range of tasks of a fully developed crew department *directly* employing sea staff in a country where legislation, traditions and standards of living demand that care and attention be paid to employees.

The range of services is not necessarily so large in all countries. The crew departments of some direct employers of sea staff are very basic. The seafarers are paid a high rate per month for the voyage only. Money for use on leave and arrangements for health insurance and any additional pensions to that provided by the government is the responsibility of the crew member. The crew department is primarily engaged in manning the ships. Many freelance seafarers accept this situation, while others prefer the feeling of belonging to a company, which necessitates the provision of many of the other services of the crew department already considered.

The other systems of employing staff will now be considered.

EMPLOYMENT OF CREW THROUGH A NATIONAL ORGANISATION

In this arrangement many, or all of the tasks undertaken by a company crew department, are carried out by a national organisation or body often with a head office in the capital and branch offices in all the major seaports. The shipowners and unions are often represented on the board of directors but the actual organisation is administered internally. It removes the responsibility of the prime task, i.e., manning the ships, from the shipowner, although he does of course have to pay for this service.

In this case the shipowner's prime task is to decide how he wants his ships manned, and in liaison with the organisation, when he wants them changed. He may pay the men through the ship's master or they may be paid through the federation.

There are advantages and disadvantages in this system: to the small shipowner not wanting the expense of a crew department it can be very useful. To the industry as a whole there is the advantage of more efficient manpower planning and distribution resulting from the overall control and knowledge of the industry's needs.

The disadvantage lies in the lack of company interest by the employees, although in some cases arrangements can be made to employ the same staff continuously within the system, provided both employer and employee are willing.

When using this system the shipowner may maintain a small "department" to co-ordinate crew matters with the employment organisation, or the work may be undertaken by one of the other departments of the company such as accounts or technical.

EMPLOYMENT OF CREW THROUGH A UNION

This is principally a North American system, although in countries where there is a closed shop agreement, i.e., where it is agreed between owner and union that all employees shall belong to the union, there is much greater union influence than if there is not a closed shop agreement.

Where the union supplies the whole crew the shipowner has little control over who sails in his ships. Otherwise the system is similar in operation to that of the industrial organisation.

EMPLOYMENT OF A CREW THROUGH AN AGENCY

This is very much like using a national organisation, as the agency provides much the same service. The principal difference is that agencies are usually used by shipowners to provide crew from countries other than the country of management or registration of the ship.

Like the national system, owners can and often do, employ the same senior and key staff from the agency in their ships, thus providing continuity and a company association with the staff.

Again, although a crew department may not be considered necessary by the shipowner, there will be a need for someone to liaise with the agency and agree conditions of service from time to time. This will require knowledge of local conditions even though the bulk of the work will be carried out by the agency.

A COMBINATION

It is not unusual in many countries and companies for ships to be crewed by a combination of the systems mentioned in this chapter. The senior and key officers are often company employees whereas the remainder are either directly employed on a voyage basis or engaged through a national organisation or agency. These combinations work reasonably well, and the size of the crew department will reflect the numbers directly employed and the functions and tasks they are required to undertake.

Budgets and Costs

Regardless as to how the crewing of ships is arranged, someone will need to provide an estimate of the costs of providing crews for each ship in the company's fleet.

In a company where all the crew are directly employed this task will fall upon the crew department. The work associated with the preparation of a crew budget is covered in *Running Costs*.

In companies which sub-contract their crew work through a national organisation or agency, the basic work to provide a cost per man may be provided, and all that needs to be added to the resultant crew cost is the company variables of overtime (if this is not a set allowance), travel costs, and overlap.

If there is no crew department the work will have to be undertaken by another department or by the accountants. Essentially, budgeting for crew costs is not difficult apart from guessing the results of future wage negotiations. It is a detailed business although it may be "packaged" by the agency to suite an owner's requirement, i.e., an inclusive cost per man per month may be agreed such that the owner is not involved in the detail of leave and other payments.

The costs of the crew are a major part of the total ship budget and for this reason the crew department, or whoever is responsible, needs to have the cost data at their fingertips, especially when wage and wage associated changes are taking place, so that the effects of such changes can be quickly assessed. In this, once again, the computer can be of considerable value.

The Control of Costs

Having produced the budget, controlling costs is centred on crew changes and standby or overlap costs, as most other costs are virtually outside the control of the department once the manning scale, establishment and training policies have been decided. Even overtime may be largely fixed, or regular and beyond the control that lies with the sea staff.

Accounts over which the department have direct control should be approved, and preferably coded before passing to the accounts department for payment.

Records

Most employers in any business keep records of staff for two reasons: because they are required to do so by the laws of the country and for their own reference purposes. Often such records need to be kept for some time after an employee leaves the company's service, as enquiries may be made later.

The essential details required by most companies are: full name, date of birth, government reference numbers (social securities, tax, etc.), next of kin, address, telephone number, and date of commencement of employment. If the employee has left the company the date of leaving and reasons for leaving.

There should then be a record of all qualifications gained and training received by the employee, giving dates, grades, etc. and a record of the employee's service. This will usually commence with the employee's application form giving details of service in other shipping companies and will then continue with the company's own record stating:

Appointments to ships and positions held.

Date of leaving ships and reasons.

Records of leave taken, sickness leave, study leave (for government examinations), time spent on company or industrial training courses, promotions and demotions and any disciplinary action.

Dates of salary changes and amounts and details of membership in any company's, or industrial pension, or superannuation scheme should also be included.

Of much importance is the care of confidential reports from ships and the technical or supplies department. Those who make such reports rely upon the crew department and owners to keep the reports confidential. These reports complete the record "picture" of each employee, such that all aspects can be reviewed when considering promotion, dismissal or any other aspects of the employee's career.

Choice of Crew

In many shipping companies, the owner or ship manager has no choice in the nationality of the crew he employs; they must be the same as that of the registry of the ship.

But others are not so restrained, although they may have difficulties in some parts of the world through outside organisations, such as the International Transport Workers Federation (ITF) and local port workers' unions who wish to protest against Flag of Convenience (FOC) Ships.

These restraints apart, if an owner is able to choose the crews he wishes, he should seek the advice of the crew department if he has one, or other knowledgeable persons, on the most suitable crews in terms of cost, training, experience and availability in relation to his ships. In this he should also seek the opinion of the technical department or others with knowledge of the particular crews.

The Staff

As has been seen, the numbers of staff employed will depend upon the system of employing crews and the number of ships employed. Like many industries, the crew or personnel function has been the poor relation of the other departments, and the staff have often been co-opted from other departments in the company because of their apparent suitability rather that any professional training in this field.

In many companies that directly employ sea staff the department manager is an ex-ship master, as his general training and knowledge of sea staff conditions of service are suitable to the job.

Other staff members may come from a variety of jobs and the main requirements seem to be good interviewing technique, and the pleasant and persuasive manner needed, for example, when trying to get someone to shorten his leave to join a ship. A knowledge of accounts requirements is usually learned on the job.

Professional personnel staff have been utilised in major companies.

Chapter Nine

Technical

The technical function of ship management is to care for the ship, the cargo and all who sail on board, within the restraints referred to in Chapter Two.

In more detail, this includes the care of all of the ship's structure and accommodation; the engines, auxiliaries, plant, cargo handling and carrying equipment; the catering and other crew support systems and all navigation, safety and anti-pollution systems and equipment.

The Activities

The activities associated with the function are carried out between the technical experts ashore and staff in ships. From time to time their work is supplemented by assistance from outside experts and special facilities such as dry docks and repair yards. Alternatively the work may be subcontracted to specialists such as safety experts, naval architects, repair and maintenance consultants, navigation and radio and radar equipment specialists and even fumigators.

Additionally, both the shore and ship staff work closely with classification and government surveyors and inspectors, whose job is to ensure that ships are maintained to required standards.

The activities of the department as a whole can be described under the following headings:

Maintenance and operation of ships and equipment to Government and classification society requirements and the owners's standards.

Formulation of work plans and budgets, and the provision of control to achieve the desired results.

Provision of adequate parts, equipment and services to ships.

Collection and distribution of documents and technical information appropriate to ships of the company's fleet.

Maintenance of records for analytical and legal purposes.

Provision of advisory and emergency services.

Co-ordination with the ship's operators and other management departments to ensure that the ship is run as safely, efficiently and economically as possible, and is maintained in service to the maximum extent required by the operators.

Supervision.

Considering these headings separately:

Maintenance

Generally speaking maintenance falls into two broad categories; necessity and efficiency.

Necessity maintenance is that which must be done to enable a ship to conform to the standards of the classification society and the government of the country of registration. If the ship is to visit countries which insist upon additional standards it will be necessary also to conform to those standards, such as the USA Anti-Pollution and Navigation rules, the Australian Cargo Gear regulations and the St Lawrence Seaway regulations.

The owner has little choice in maintaining his ships to these standards. Although there is sometimes flexibility in the classification society standards, any allowance is usually based on the good maintenance record of the ship and owner.

The second category concerns the efficiency of the ship. This is achieved through expert maintenance and control of the hull, equipment machinery plant such that minimum fuel, lubricants, and parts are used and overhauls and breakdowns are minimised.

There is, of course, overlap between these two categories and a fuel efficient engine and underwater hull section will probably conform equally to classification and other standards. But the hull structure can conform to those required standards and yet not be sufficiently smooth to gain those extra points of fuel savings. This applies also to other plant and equipment, i.e., the standards are for safety and not necessarily for efficiency.

One of the major policy decisions a shipowner has to make is which of these categories is more suitable to his type of ownership or whether he requires the ships maintained at some level in between. This decision will depend on his policy of buying or selling ships, which in turn will probably reflect the funds or credit he has available.

To give examples of the types of policy and the background often associated with them:

The new ship: This can be relatively cheap or expensive. The cheap ship will be according to a basic specification which may, for example, have basic internal and external coatings of tanks, decks, and hull, etc. Such a ship will require attention from the beginning, with gradually increasing costs for scaling, repairing and recoating the surfaces. Alternatively the owner may decide on a policy of minimum maintenance from the onset with a view to selling the ship after five years, thus leaving some other owners with the problems of lack of maintenance.

The relatively expensive ship is one in which the initial expenditure is high as a result of additions to the basic specification, such as high quality coatings and other methods of structural protection such as cathodic protection. The owner who pays for these extras will probably have a policy of retaining the ship in his service for at least ten years, such that the initial extra expenditure spread over the ship's life will be less costly than maintaining a basic specification ship as described above.

There would, of course, be some maintenance requirements for the well protected surfaces and although not cheap, in the long term the overall costs would be less as there would be much less damage to the structure through corrosion.

In the unfortunate event that the owner of such a ship has to dispose of it before the end of its anticipated service life, he would hope to obtain a relatively better price for it than a ship of the same age, built to a basic specification with subsequent minimum maintenance.

The second-hand ship: As has been seen, this can be of the well maintained or poorly maintained variety. Although it may not seem logical, there are owners who deliberately buy somewhat run-down ships because they can get them more cheaply than well maintained ships. With cheap crews and cheap repair facilities they believe they can return the ship to an acceptable standard at a total cost below that of buying a well maintained ship.

The policy after purchase will depend upon the owners trading philosophy. He may only trade the ship for a few years and then sell it again, carrying out minimum "necessity" maintenance in the interim.

These are the broad philosophies behind maintenance policies and although examples of hull and structure maintenance have been given, such policies can apply in the same way to machinery, equipment and plant. Essentially the policy is a market decision. The owner gets what he pays for and the condition in which he maintains his property depends upon the "necessity" factor and how he intends to run and eventually dispose of the ship.

But as in all matters in shipping, owners need to be flexible in their policies and although they may have a defined policy, that policy may have to be adjusted from time to time as affected by market forces, e.g. a company with a policy of continuous, planned maintenance to a high standard may have to reduce that standard at times of low freight rates. Similarly a shipowner with a policy of minimal maintenance and short ownership may find himself in the position of wanting to retain the ship and thus having to spend more than intended on necessary maintenance.

As was seen in Chapter Two, the technical manager and his staff, both ashore and in the ships, are restrained or must work within these owner's policies. They are also restrained in their tasks by the classification society rules, government and international legislation, budgets and funds available, and the morals and ethics of the company.

Within these restraints the manager should maintain the ship by:

Planning, budgeting, and controlling

The factors involved in planning and budgeting for the technical department have been described in some detail in *Running Costs*. Briefly, the managers concerned must decide what they intend to do during the financial year before they can prepare plans and thus their budget, for submission to senior management.

The more the maintenance is planned and the more information available about the state of the ship and its operation, the less guesswork there will be in the plan of the work and the eventual budget.

At this point it is necessary to clarify what is meant by "planned maintenance".

Planned maintenance is a system of controlling the maintenance of a ship through schedules of regular examinations, overhauls, renewals or replacements, coatings, etc. of all parts of the structure, machinery, equipment, etc., *with flexibility*.

This means that every part of the ship and every piece of machinery, plant and equipment and system are listed and given a time for periodic examinations and/or, overhaul, recoating, etc. The periods are based on running hours in the case of machinery, or months or years for other parts. They are formulated to coincide with classification society and government requirements, the manufacturers recommendations, and the experience and records of the owner's technical advisors.

Flexibility is built into the system, when allowed by the classification society or government rules, to avoid unnecessary maintenance when machinery, in particular, is running well. The decision to extend, or if necessary shorten the period between scheduled overhauls, etc. is usually made on the basis of visual examinations, gaugings, and the use of performance monitoring equipment. This equipment aids an expert in its use in assessing the wear and efficiency of machinery while running, i.e., without the need to take it apart.

It is noteworthy that without such flexibility, planned maintenance itself can be expensive as a result of over or unnecessary maintenance. Without planned maintenance and associated condition and performance monitoring and fault diagnosis, such maintenance as is done may not be adequate and there may well be a number of breakdowns with high costs, and out of service time involved. Alternatively, some maintenance may be excessive and create problems by disturbing well running machinery, apart from the costs of the manpower and spares involved.

Similarly, without periodic inspections and examinations of parts, the owner will not be able to predict with any accuracy the costs of surveys when due, because he will not know the extent of any renewals or replacement parts which may be required.

Thus the quality of planning and budgeting in the technical department will vary directly with the maintenance policy of the owner.

In the same way the controls and control will depend upon that policy. The quality of technical control depends upon the quality of the information received about the maintenance and the performance of the ships, as referred to in Chapter Five. It also depends upon the quality of the staff, be they in ships or ashore, who must collect, analyse and act on the information.

If a shipowner has a management control policy, the setting up of planned maintenance and performance controls systems and the training of staff in their use, can be expensive initially. But the long term gains through greater efficiency, reduction of out of service time, repair and replacement costs should outweigh that cost. Provided always that the controls themselves are efficient, and are in line with the guidelines given in Chapter Five.

The advantage of these controls is that they give the manager and those with whom he works, the opportunity to decide which is the best action. For example, if the planned maintenance system indicates that there is, or is going to be, a shortfall in the work plan then corrective action may need to be taken. This can be by the use of extra sea staff of specific categories, a squad of specialists placed on board to carry out specific work at sea and in port, or shore repair yard staff and facilities.

Without such controls systems the manager has only limited knowledge about the performance of the ship and work being done, thus any control action he can take is limited, and the potential for unexpected expenditure, high. For this reason there is usually a need for much on-board supervision by the shore staff, particularly when the crew are not regular employees or not highly trained.

But control of technical work and costs is not confined to the work of the ship's staff and the performance of the ship. Some of the highest costs and time out of service are incurred while the ship is undergoing major repairs involving shore facilities, such as dry docking, modifications and repairs beyond the capacity of the ship and its staff.

In this the essence of control lies firstly in defining the work as completely as possible, and obtaining the best tender for the work in terms of cost, time, workmanship and geographical position. Most importantly the operator's trading requirements must be taken into consideration. When the decision has been taken as to place and contractor, the control lies in supervising the work done, ensuring that "extras" are minimised and finally in checking and negotiating the accounts.

This section has been written in the context of the technical manager and the shore staff planning, budgeting and controlling. As mentioned in Chapter Three, with an SDC organisation much of this work will be carried out by the sea staff.

They will, of course, work in conjunction with the shore staff from whom they will seek information, advice and comments. In the same way that the shore staff will turn to the sea staff in a company operating a centralised ship group system with regular sea staff and good management systems.

As mentioned in Chapter Three, there are some technical areas beyond the scope of the sea staff, because of the nature of the search for the optimum repair yard and specialist knowledge required in drawing up specifications and evaluating tenders.

But whenever possible in an SDC organisation the system will be geared for the sea staff to make decisions for the areas of costs for which they are accountable. Thus in the case of a shortfall in the planned maintenance schedule they will decide how best to make it up, in consultation with the shore staff.

Provision of spare parts, equipment and services

Spare parts or spares: This is covered in some detail in *Running Costs*. Briefly, spares are an item requiring careful control. Too many is cash lying idle, while shortage of the right spare when required, can involve the ship in considerable expense through delay and out of service time.

Ideally a spare gear system should be set up when the ship is new. The longer the installation of the system is left, the more difficult it becomes to arrange, as used and partially used parts become mixed with new parts, and serial numbers and sizes are mislaid.

The system should be based on a standard level of spares dictated by classification society requirements, manufacturers recommendations, owners experience and an assessment of the likelihood of some parts being required and the delays which may occur if they are not available.

Essentially all that is required is a system of recording and notification when parts are used, so that replacements can be ordered and transport arranged. Regardless of how the company is organised the chain of events after a spare part has been used starts at the ship and is then dealt with ashore, either by a member or section of a traditional technical department, a technical service department, or even a consultancy organisation specialising in the supply of spares.

The great problem with spares is the length of time between ordering and delivery. Some spares, such as cylinder liners and pistons can be very expensive, therefore the engine builders only keep limited stocks. If there is a sudden demand for a particular spare it means that more have to be manufactured and so a delay in a delivery may occur. For this reason, some shipowners with a number of ships with similar engines or equipment, "stockpile" some spares at strategic points around the world to suit the trading patterns of their ships. This should only be done after assessing the advantages and disadvantages of keeping a stockpile, as large sums of money can be tied up in this way.

Equipment: Generally speaking tools, lifting gear, materials, etc., are provided by the supplies or purchasing department in consulatation with the technical department, and this is discussed in the next chapter. In small companies the technical manager may arrange for these items directly.

In general anything that assists the sea staff in carrying out their operational and maintenance work is of value, providing that the anticipated usage justifies its cost, and most importantly, the staff are fully trained in its use and are able to interpret the information provided. This is particularly so in the case of sophisticated equipment as used in performance monitoring.

Services: The technical function of the marine industry covers a wide area and it is inevitable that there are specialists in certain sections, ranging from cathodic protection to vibration and even micro-organisms in lubricating oil systems. From time to time it will be recognised that there is a particular problem beyond the knowledge or experience of the ship and shore staff, such that an expert needs to be called in. In some cases the manufacturers of equipment and supplies will provide such a service free as part of the back up to their product. In others the service has to be paid for. In this, the wisdom of the technical staff will lie in the recognition that they have a problem to which a solution may be found through someone with more specialised knowledge than themselves.

Documentation and information

Documentation: A most important task of the technical department is to ensure that ships have the appropriate certificates to allow them to sail. Although the various surveys have been carried out to ensure that the ship's maintenance does conform to the government and classification society rules, responsibility lies with the department, in association with the ship's staff to ensure that all the appropriate certificates are on board each ship.

Similarly they must ensure that the ships have all plans and diagrams as required by law and all appropriate operating manuals. The majority of these will have been placed on board when the ship was delivered from the builders, but a constant check needs to be maintained to ensure that they are on board. Similarly, when modifications are made, the department must ensure that the plans, etc. are amended appropriately.

A company's technical manual giving additional machinery and equipment operating advice and instruction, may be included with these documents.

As mentioned in Chapter Four, the company's regulations should require a list of documents to agree when senior officers change over.

Information: In any industry there is a need for staff to keep themselves up to date on technical developments, hazards, safe practices, etc. in their particular field. Nowadays there are so many technical developments in the marine industry and so much information published by governments, industrial and professional bodies and journals, that it can be difficult for staff to keep abreast of it all.

By the nature of his stable position ashore, the technical manager usually has the responsibility of gathering this information and deciding how it should be used. It usually falls into two broad categories: development, and essential information:

Development information: This is information which could be used if modifications are contemplated for existing ships or when new tonnage is being considered. Essentially it is reference material and should be retained and filed as appropriate to the type of ship operation in which the owner is engaged. As with all information, care needs to be taken in selection as it can be a time consuming exercise and take up a lot of space. Before storing such information the manager should consider whether or not it is readily available elsewhere, such as in professional institutions, government officers or in colleges.

Generally speaking, this information should not be issued to ships, although enlightened owners with stable sea staff may consider supplying ships with technical journals to maintain professional interest and development of the sea staff.

Essential information: This usually concerns the immediate safety of the ship, crew, cargo, and the protection of the environment. The source of much of this information is usually the government of the country of registry; international and national institutions such as the IMO, ICS and the General Council of British Shipping. It usually takes the form of new legislation or official warnings of hazards, dangerous operational practices, anti-pollution requirements and such like. In some cases the information may be educational, in the form of posters, booklets or case studies of accidents.

The company may be given this information or should be aware of its existence and arrange for it to be supplied. It is usually supplied by government publications offices or departments, industrial federations or forums, and P and I Clubs. In the main, the role of the technical manager will be that of distributor, withholding only that knowledge which is not relevant, e.g. not sending information about the carriage of dangerous chemicals in bulk to a refrigerated cargo ship. The manager may feel it necessary to draw the ship's officers attention to some of this information and may even wish to ensure that it is officially retained on board the ships as has been considered in Chapter Four.

Some essential information may come from experiences within the company and the manager may consider it necessary to give guidance or instruction on some of it as has been discussed in Chapter Four.

Other essential content is of the "aid" type and covers such matters as up to date port information not always covered in official guides and charts. Again the technical manager will need to be selective in what is issued to a ship, as there is little point in giving information on ports unlikely to be visited.

In other words, care needs to be taken that only relevant information is issued to ships.

Maintaining records

As has been described in Chapters Five and Six, record keeping is important. Technical records are often a key factor in inquiries and in refuting claims against ships and owners. Other uses are in the analysis of a ship's performance and its equipment and the preparation of budgets.

Records are usually kept in the technical department of the following:

Copies of ship's deck and engine logs.

Performance data such as speeds, consumptions, past cargoes carried, also distribution and stability data.

Ship and machinery condition and maintenance records, which includes reports on repairs and modifications, dry dockings, etc.

Survey and inspection reports.

Lists of equipment and spare parts.

Copies of all certificates issued to ships such as S.E.C., SAFCON, MARPOL, Loadline, the Safety Radio Certificate, and the Ship's Register.

Copies of all plans, operating manuals and technical instructions issued to ships.

Costs information from repairers and suppliers and past accounts.

Records of insurance claims including heavy weather and other damage, and associated survey reports and repairs.

The computer lends itself to much of this record keeping and subsequent analysis as described in Chapter Six, particularly with planned maintenance and spare gear and supplies control systems.

Provision of advisory and emergency services

The technical department usually has responsibility for providing advice and recommendations to the board of directors on all important company regulations, policies, and instructions affecting ships. The advice and comments of sea staff may also be taken into consideration, particularly in a company with regular employees at a senior level.

Such advice and recommendations will usually be in the following areas:

Formulation of new regulations, and instructions to conform with new legislation and company's standards or policy.

The manning scales for each of the company's ships in consideration of the relevant government regulations.

The training and experience of national and foreign crews.

Supplementary training requirements for sea and shore staff as referred to in Chapter Eight.

Disciplinary measures resulting from company and other inquiries into accidents and behaviour of staff. (See also Chapter Eight.)

Modifications to existing tonnage to conform to new or anticipated legislation, or the operator's changing requirements.

Specifications for new tonnage to produce the best possible ship for the operator's foreseeable requirements, foreseeable legislation, efficiency and cost effectiveness.

Recommendations on the purchase of second-hand tonnage including the costs of bringing the ship up to the owner's standards.

Fortunately, emergency services are rarely required, but the fact that when they do occur they are unexpected and need immediate action, necessitates clear procedures both ashore and in ships. This will include details of staff on call at out of office hours and the facility of access to records and communications at any time. The names, addresses, telephone and telex numbers of staff agencies, official organisations, coastal radio stations, repair and salvage organisations, etc. should also be easily accessible.

In this there should be procedural instructions issued to the ships, so that the Master can be clear on how he seeks advice, e.g. during out of office hours and what he should do in respect of such matters as salvage. This is referred to in Chapter Four.

Similarly, staff should know what is expected of them in ships and should have adequate training and organisation to meet such emergencies. It is the responsibility of the technical department as a whole, i.e. ships and shore, to ensure that such training and organisation exist.

Although most emergencies are essentially of a technical nature, e.g. stranding, collision, fire, oil pollution, etc., there is usually involvement with other departments, such as the insurance department who may need to inform the underwriters and P & I Club, the personnel department who may be in contact with next of kin, and the operators, who may need to be in contact with the cargo owners. For this reason, there is a need to designate someone to be in charge of the shore organisation during the emergency. In a ship group organisation this will probably be the Ship Group Manager, in a large centralised organisation, it will probably be the fleet manager. Although it is important to select someone who already has a co-ordinating function, the most important factor is for everyone to be clear on who is in charge.

Co-ordination with the operators and other management departments

Regardless of the type of organisation, there is a need for co-ordination between those who carry out the technical function and the operators. Although the crew and supplies departments must also co-ordinate with the operators, most of their requirements can be arranged without interrupting the ship's trading. This is not always possible with technical matters and although there can be flexibility between the operating and technical requirements, there are other times when the ship has to be taken out of service in order to be maintained to the various requirements referred to at the beginning of the Chapter.

In an SDC type of organisation, or one of dual accountability such co-ordination will be through the ship group manager or fleet manager. In other organisations there may be direct co-ordination between the technical manager and the operators.

The importance lies in each recognising the needs of the other while considering the effect on the company as a whole. There is little point in sending a ship to a cheap dry dock, whose costs will maintain the technical budget for the ship, if the loss of earnings off-sets this advantage. In the same way, sending a ship to an expensive dry dock because it is very close to a port where high paying freight is available, should only be considered if the additional earnings out-weigh the extra costs.

Co-ordination with the crew department involves consultation on training, assistance in disciplinary matters, and authority for appointments and promotions mentioned in Chapter Three.

As will be seen, co-ordination with the supplies department involves approval of supplies and equipment when necessary.

Co-ordination with the insurance department involves the collection of data for claims negotiations or legal action. When damage to the ship is incurred, the department works closely with the insurance and classification society surveyors, in conjunction with the insurance department, before putting repair work in hand.

Co-ordination with the accounts department involves the approval, and preferably, coding of accounts. In addition to bugeting, the department will have to provide the accounts department with estimates of large items of expenditure for anticipated work, parts, and equipment.

Supervision

Reference has been made to supervision of the sea staff in Chapter Three, and its importance in relation to the owner will be considered in Chapter Twelve. Although part of the control activities of the department, its importance is such as to warrant further mention.

Although supervision is carried out by the staff of other functional departments and senior management, by visits to ships, the weight of this responsibility in the main, falls upon the shoulders of the technical staff. This is principally because they are best fitted to judge from what they see and from conversations with the staff, whether or not the ship and staff are as they should be.

They will often combine such visits with the supervision of difficult or major repairs involving shore repairers, or experts, for specific parts of the plant or machinery.

Supervision is also carried out by examination of log books, report forms, and other data as described in Chapter Five. Although the superintendent has to be on the look out for reports which record events which may not have happened, for instance boat and fire drills; in the main such reports do give a good indication of how the ship is being run, in addition to physical checks.

Staff

At the beginning of the 1950s many of the traditional shipping companies had almost as many marine superintendents as engineer superintendents. But with the addition of experts in electrical and electronic matters, the engineer department increased considerably in comparison with the marine superintendent's department.

The next step in many companies was the merging of the marine and engineer superintendent's departments into a new technical department. With this came the transfer of the "deck" maintenance responsibility from the marine to the engineer superintendents who became known as maintenance superintendents. Thus there was a reduction in role of the marine superintendent and a consequent reduction in their numbers.

Today, in many of the more developed shipping company organisations, the marine superintendent or person of similar title such as marine manager, marine inspector or port captain, has become an expert rather than a broad based superintendent. His work is now principally associated with navigational, safety, and anti-pollution matters. When required, he provides advice on the carriage of specific cargoes and operational problems, and is available to enquire into accidents associated with the navigation of the ship and the carriage of cargo. His background is of sea service, culminating in a number of years experienced in command of ships. It is usual for him to be promoted from the sea staff of the company although some companies do recruit from outside.

As noted, the role of the maintenance superintendent includes all maintenance, although in large companies specialists may be appointed to deal with areas of electrical and electronic matters and there may be a separate "development" section formed, to deal with modifications and new tonnage and even major repair work.

Whereas the background of the traditional engineer superintendent was similar to that of the marine superintendent, i.e. sea service to chief engineer in the same company, this has changed. Superintendents are still promoted from the sea staff, but there has been an increase in selection from other marine associated backgrounds, such as repair yard manager or naval architect. In the electrical and electronic fields recruitment may be from the services or other specialised shore industries.

The organisation of the staff has been considered in Chapter Three. In a traditional organisation the technical department is a whole unit. In an SDC organisation, or variations of this theme, there can be a maintenance superintendent attached to the ship group manager's team, while separate units or departments of expertise deal with such matters as major dry dockings and repairs, safety, and anti-pollution, the issue of technical information and the formulation of regulations, etc. for the fleet as a whole.

Like other organisations, the complexity of the department depends to a large degree on the size of the fleet. Whichever way it is organised, any shipping company must have expert knowledge available on every technical aspect of the ships they own and operate.

But the way in which this expertise is provided depends upon the owner's philosophy of ship management. Much of the work can be given to outsiders who can do everything from inspecting ships, to providing spare gear control and planned maintenance systems, oversee dry dockings, approve accounts, arrange spare parts and provide a complete nautical publications service. A relatively new development has been the contracting of the whole engine maintenance to an engine builder, as in the maintenance of automation equipment.

In a small one or two ship company, utilisation of outside expertise would seem a sound policy. A large company may wish to consider having its own experts, although there is a tendency for such staff to grow unless strict control is exercised.

Chapter Ten

Supplies

"I know where you can get it cheaper"
(a common expression used by people who usually cannot)

The function of the supplies department of ship management is to ensure that each ship has sufficient equipment and stores, in order that a shortage of any item will not hinder the progress of the voyage, or cause a hazard to the ship, or those who sail in the ship, or to the cargo. As will be seen, "supplies" covers a large number of items and whereas there are some items the ship can manage without for a period of time, such as a paint brush, there are other items, such as up to date distress flares or a fire hose, that the ship cannot be without. Not only would the ship be in danger if it did not have the right firefighting equipment, but a local government inspector may refuse to allow the ship to sail without it.

Before describing the activities of the supplies department it is preferable to define what is meant by equipment and stores.

Equipment

When a ship is built the builder's specification will include a number of items to "outfit" the ship. Apart from the basic articles of moveable equipment essential for the ship to be able to obtain the requisite certificates (such as flags, lifebuoys, ropes and wires to work cargo and secure itself to a quay), little else is supplied unless requested by the owners.

Charts and publications, nautical instruments, tools, repair and maintenance equipment such as a paint spraying machine, hammers and washdeck hoses are usually supplied by the owner.

Some of this equipment may last the whole life of the ship or only need to be replaced because it has been lost, stolen, or broken. Other equipment, like washdeck hoses and cargo blocks or even carpets in frequent use, do wear out through time and have to be replaced.

If the trade of a ship does not change significantly, i.e. it continues to trade as designed, this "stock" of equipment provided by both builder and owner does not change very much during the ship's life. But it will need to be listed, maintained and checked from time to time.

Stores or consumables

These traditionally fall into two sub-categories and as they are often dealt with quite differently it is as well to look at them separately:

Victuals or food: The supply of food for the crew is arranged in three basic ways:

In one the crew are paid an allowance with which they buy the food through the chief steward, cook or one of their members. The only "supply" by the owner is the cooking equipment and refrigerators and store rooms in which the food is kept. It is noteworthy that in some ships the officers' food is supplied by the owner and managed by the chief steward or cook steward, while the ratings buy their own food with an allowance.

The crew allowance system can be very convenient to the owner because it completely removes an area of responsibility, and thus the need for shore support.

In another, the chief steward or cook has a contract with the owners by which he is paid an agreed sum per head, per day, to supply the crew's food, plus additional allowances in some climates and when shore visitors are provided with meals.

In the third way the owner supplies the food. In essence this is the same as the contract with the chief steward or cook, as the owner has to budget, and so has a target figure for the cost per head, per day. The principal difference is that the owner has to provide a shore organisation to support the chief steward.

With both these systems there are the restraints of any government victualling scales to consider, and the owner's policy; which for example may be to provide fresh or frozen fish three times per week and meat for two meals each day for the other days. Such a policy is particularly necessary when an owner has a large number of ships and staff moving from ship to ship and consequently make comparisons of the standards of meals in each ship.

For the purpose of the remainder of this Chapter it will be assumed that the owner supplies the food.

Other stores: This covers all other consumables such as paints, greases, cleaning materials, packing, ropes, wires, and chemicals. Lubricating oil is usually arranged through the supplies department but may be arranged through the technical department. Similarly paints, particularly for dry dock use, may be arranged directly by the technical department.

In the main, supply of the bulk of the stores is arranged through the head office with the sea staff "topping up" or completely replenishing stock as required. Like victualling, storing can be contracted out, but as will be seen later, there is need for co-ordination with the technical department.

Supplying Ships in Practice

The factors

Secondary only to the prime function of ensuring that ships have sufficient equipment and stores, is the function of seeing that the ships are supplied *economically*. This means that not only should equipment and stores of the right quality be purchased at the lowest possible price, but that no more stores than are economically necessary are kept in stock. Stock is money tied up and unless there are sound reasons for keeping large stocks in ships, such as unavailability or high cost of a commodity in an area in which the ship is trading, they should be kept to a minimum. Even when there are sound economic reasons for keeping large stocks of some commodities, due regard should be taken of its shelf life, i.e. how long it can be kept without deteriorating.

In the same way consideration must be given to the actual storage space in the ship. It may be economically sound to place a year's supply of paint or lubricating oil in a ship expected to trade for a long period between expensive ports or ports where stocks are unobtainable. But if the ship has only space for six months supply, there is little that can be done.

So firstly, the supplies manager and his staff must know what equipment and stores each ship has to carry and what it consumes. They must then decide the maximum and minimum stock levels of each commodity in relation to its shelf life, such that supplies strategy can be planned. That is, they can decide when major storing should take place and where and how much should be put on board the ship.

As will be appreciated, the number of items and sizes of items required for a ship runs into thousands. It is little comfort to the supplies department to be told that nowadays there are far fewer items than in the ships of the 1950s. Work study, design, and new materials have made so many of the old stores items obsolete (holy stones, brass polish, coir matting, pitch for the wood deck seams and canvas.)

Nevertheless, supplying ships involves considerable "detail" and this in itself should be controlled. Having identified the items required to run the ship, the items themselves should not be allowed to increase without very careful consideration. Similarly the types and specifications of items need control and in the case of anything likely to affect the safety or maintenance of the ship, the technical department must be consulted. This is particularly so if the item is of a chemical nature likely to be incompatible with other materials.

Having gathered all this information, the manager is in a position to arrange inventory and control systems, to which the computer ideally lends itself. This can be arranged into groups for convenience and budget purposes as described in *Running Costs*, and coded and subcoded for easy entry into the system. Thus all paints will be in one group, all cleaning materials in another, all ropes and wires in another and so on.

The next step is to ascertain the average consumption of each item so that a "norm", or target, can be established against which actual consumptions can be compared. This target can be adjusted from time to time in the light of experience. Additionally, allowance should be made for high consumptions of some commodities at special times such as paint in dry dock. In this the supplies manager will use his own experience and that of his staff, the advice of the sea staff and the technical department.

Following this, a system of control must be arranged so that stock levels are always known within reasonable limits, and consumptions and replenishments are communicated as necessary, i.e. in time to take action. For most items a three-monthly report may be adequate but times of high consumption may need to be reported more frequently, say on a monthly basis. At the other extreme "equipment" can be quite separate from stores and need only be checked or reported at six-monthly intervals.

The format of the control information should also show the maximum and minimum stock levels within the restraints of shelf life and space already mentioned.

The sea staff will need to be instructed on these requirements and the method of communicating the details decided as outlined in Chapter Four. As also shown in that Chapter, from time to time physical checks will need to be made of the stores on board, to ensure that stocks are as recorded.

Thus the supplies manager needs to know the commodities, quantities, consumptions, minimum and maximum stocks for each ship. He needs this in a concise, easily readable form, or on a computer screen with print out facilities. But the sea staff who are working with these commodities must produce the data and work with it too. For this reason, as mentioned in Chapter Four, any system devised must be workable for them too, so that control can be exercised in the ship first.

Having decided what is wanted the next step is to decide how it should best be provided:

Specification and cost

There are two aspects of purchasing which although generally applicable to all businesses, are critical in ships. They are the specification and the price.

Specification is the first consideration, because if the stores or equipment are not of the right size and quality there is potential for loss or danger such as in the case of lubricating oils and cleansing materials. Even such small items as torch batteries can create problems if the stock is calculated on high quality batteries, and a cheap type is supplied with only half the life.

Thus a shipping company needs to have some "standards" for its supplies and should only deviate from them when the supplies they require are unavailable. In many areas it is penny wise and pound foolish to buy cheap stores, while in other areas a cheap commodity will serve its required purpose just as well as an expensive one. With some commodities the supplies department staff will be well experienced in these matters, but in others as already stated, the technical staff should be consulted.

Whenever possible items should be standardised, i.e. the minimum number of sizes, as with screws, nuts and bolts, and types, as with lubricating oils, should be supplied to ships, and whenever possible sizes and types should be those which are most easily obtained internationally. Some of these requirements may start in the design specification of the ship, while others require a search for a common alternative as in the case of lubricating oils. For example, the manufacturers of a number of pieces of special equipment may each specify a different lubricant or grease when, on investigation, it could be found that one lubricant or grease is suitable to all, thus facilitating the ordering and supply of these items.

Having decided on the specification and type, the next thing is to decide how and when to purchase and supply.

Purchase and supply

Compared with similar departments in other industries, this function of the supplies department has the added difficulty that the ship is constantly on the move. In many cases the pattern of movement is so irregular and uncertain, that the development of any supply strategy is very difficult. For this reason much planning of tramp shipping supplies is often opportunist rather than planned from a long term point of view. For ships on regular line trades the scheduling can be easier, but may be more expensive, or cheaper, depending upon the prices and availability of various commodities.

Thus a knowledge of world prices and availability of stores is essential to supplies department staff. In the past they would know when to supply commodities on the basis of their experience, but today the computer can aid them if such prices and the availability at various ports are included in its program. Thus if a ship is suddenly ordered to a particular port, the information as to its stock position, storage capacity, prices and availability at the next port, can be compared immediately with the next anticipated port and a decision taken.

One of the great levers for any purchasing manager is the quantities of each commodity he expects to buy. The more he can guarantee to buy from a particular manufacturer or general supplier over a given period of time, the better his changes of obtaining a discount on the normal buying price. The irregular movement of the ship may prevent this, but there are some items of marine equipment where discounts can still be arranged. This can be so in the case of the major manufacturers of marine equipment and stores, such as paint and lubricating oil, who stock pile or arrange local manufacture under license, in all major ports. For this reason they can offer the supply of their particular commodity at a competitive price "world wide", although with some variations in discounts for different parts.

In the same way, larger shipping companies are able to stockpile equipment and commodities themselves in areas where they are difficult or expensive to obtain. In such they need to take care that the price of the stockpile maintenance and the transportation does not increase the prices to above locally obtained stores, provided they are obtainable. Alternatively they can despatch large shipments of supplies by other ships, but for this to be effective the ship's schedule needs to be firm, so this arrangement is usually only suitable for ships engaged in regular trades.

From the foregoing it can be seen that there are a number of ways to purchase and supply equipment and stores to ships. Each ship and trade will have different requirements and the only way the supplies department can effectively do the job is by having up to date information, i.e. by knowing the stock situation in the ships and the availability of commodities in various parts of the world.

They can arrange all the supplies from the head office based on this information, or can arrange major supplies while giving authority to the ship's staff to supplement them as necessary. Alternatively they can provide the ship's staff with all the information available and give them the responsibility for keeping the ship supplied. They would arrange major supply contracts when this is advantageous and assist with arrangements for obtaining any supplies when required.

Budget and costs

In this the supplies department is no different to the crew and technical departments. Plans have to be made based on assumptions and experience, and changed into money terms for the budget. Thereafter the system of controls should provide the feedback in terms of quantities and costs, such that the reasons for any variations can be identified. Again the computer can assist in this identification of variations.

The supplies function of ship management probably generates more invoices than any other and the system for the processing of order forms, receipts, and invoices needs to be carefully organised. This has been dealt with in some detail in *Running Costs*. Briefly, a standard order form should always be used and sequentially numbered to enable checks on missing orders to be made. Forms should be made in sets, preferably multicoloured, so that the ship and supplies department always have a copy and that each knows it has been made. This can also be included in the computer program.

The supplier will also have copies and will use one for a receipt when the supplies are delivered, with comments if any items are unobtainable. This should be attached to the invoice which will be authorised and coded by the ship and/or the supplies department ashore. It will then be passed to the accounts department for payment and inclusion in the management information system.

It is noteworthy that accounts for supplies are often settled by the company's agents abroad, and in this, exchange rates warrant careful attention.

Before leaving this section, it is important to note that the accountancy treatment of certain areas of costs, demands that the stock remaining on board of high cost items such as lubricating oil, be reported at set periods, usually quarterly. The cost can then be credited to the period which has ended, and then carried over as a debit to the next period, thus giving a true cost of the consumption of the commodity. A good stock control system should provide this without difficulty.

Co-ordination with other departments

As mentioned in Chapter Three, the supplies department may be considered as a "service" department, supporting all the ships in the fleet regardless of ship groups or the line management organisation of the company. But this does not materially change the way in which it carries out its function. In this it must co-ordinate with the other departments of the company, including operations, particularly if a delay in supply is likely to affect them. Depending upon the organisation, they may work directly with the department concerned or through a co-ordinator or ship group manager. As already mentioned, they will co-ordinate with the technical department wherever a specialist decision is required on products being considered for supply.

As noted in Chapter Eight, they will be required to work with the crew department in matters of recruitment, promotion, and discipline.

Advice and information

Like the crew and technical departments, the supplies department has a responsibility to advise the sea staff on the supplies the ships carry, although in the case of some dangerous products and equipment this may be given by the technical department.

The major areas on which advice is usually given are:

The care, preparation, and use of foodstuffs and other consumables.

Commodity availability and information as to prices.

Ports where regulations are in force regarding the use of certain foods in port, and the disposal of food waste.

Company contracts with manufacturers and suppliers.

Victualling scales and consumption guidelines for other consumables.

Certification requirements of certain supplies items.

The scale of this information will usually depend upon the size of the company and as has been seen before, will probably tend to follow the rule that the larger the organisation the more information it produces.

Sub-contracting

Just as the victualling part of the supplies function can be sub-contracted, so it is possible to sub-contract the whole function to a specialist supplier, who by his larger purchasing power, may be able to do the job more cheaply, particularly considering the cost of staff and office space ashore. This alternative should be considered by any shipping company, particularly one with only a few ships.

The slop chest or bond

This is a facility for the crew arranged by some larger companies, or left to the master or the crew themselves to arrange. For a company trying to reduce or contain its shore activities it is probably best to leave the matter in the hands of the sea staff.

The staff

The background of the supplies staff can vary. They may have been at sea as chief stewards, or have been employed elsewhere in the catering or supplies industries, or have developed within the supplies department from an early age.

There are no particular qualifications in this field, but knowledge acquired from years of experience in the supplies section of the industry is important. This needs to be backed by a knowledge of accounts procedures and the usual management disciplines of planning, controlling and co-ordinating.

Chapter Eleven

Insurance

Because of its close association with major financial and legal decisions of the shipping company, the insurance department usually lies outside the direct control of the ship manager. But as will be seen, there has to be a close working relationship with the ship management departments and for this reason the insurance costs are often included in the ship management department budgets.

The function of the department is to ensure that the shipowner is financially protected by insurance of his ships against;

physical loss or damage,
liability to third parties,
loss or interruption of earnings,

within the usual restraints of management, i.e. the owners policy and international statutes, governmental rules and contractual requirements. Beyond this the owner can decide the amount of additional "cover" he requires and can afford.

The activities of the department in fulfilling this function fall into the following categories:

Arranging the insurance.

Processing claims.

Providing an internal advisory service to senior and middle management.

Arranging the insurance

This is usually an annual event which culminates in the acceptance of the insurance by the underwriters and the P & I Clubs after two or three months of investigation and negotiation. The negotiation is carried out through a broker in the case of most hull and machinery insurances, and usually through a broker for the insurance of liabilities, loss of earnings, strikes and war risks.

The basic factors in the procedures involved in obtaining the insurance and its costs have been considered in *Running Costs*. As was stated in that book, Insurance is a very complex subject and for a greater understanding reference should be made to more comprehensive books on the subject.

For the purpose of this book it is sufficient to highlight the department's responsibilities in carrying out the function of arranging the insurance.

Insurance is very much a business of contracts which always need careful attention. Like charter parties, many of the standard insurance policy clauses have been tested in the courts of the older maritime nations, but new clauses are always an uncertain factor until put to the test, as are new policies.

For this reason the insurance manager and his staff need to have an intimate knowledge of all the clauses contained in the policies covering the owners' insurance. This is particularly necessary when the owners' insurance is "spread" for reasons of economy amongst different insurance markets. In such cases great care is necessary to ensure that something is not missed by spreading the cover.

The insurance department needs to ensure that all legal and contractual requirements are covered by insurance. These will include requirements of mortgagees and the requirements of governments in regard to liability for oil pollution and in some cases death and accident compensation for the crew.

In addition to the insurance cover which has to be taken, they will need to consider all other risks and weigh the costs of insuring wholly or partially against them, or the likely losses which may be incurred if there is no insurance, or if the insurance is insufficient. In this the value placed on the ship, if not decreed by the mortgagees, can be a flexible factor but the underwriters may well insist on a value compatible with repair costs up to the Constructive Total Loss Value, or may insist upon a restricted policy, if a low premium is sought, thus making a risk decision necessary.

A way of reducing premiums is to increase the "deductible" or amount payable by the owner before insurance payments are made. Again this is a "risk" decision although the risk can sometimes be lessened by insuring the deductible at a lower rate than is applicable to the major part of the insurance.

The amount of choice in this may also be restricted, as although the owner may prefer a low deductible, his insurance record may be such that the underwriters will insist on a large deductible.

In normal times most shipping companies will make a policy decision in such matters as the valuation and the deductible. Similarly they will decide on the amount of protection they require for other insurances such as loss of earnings, war risks and strike insurances.

Because of the large sums involved and often contractual obligations, all decisions regarding the amount of insurance cover will usually be referred to the senior management before a final decision is taken. In this the insurance manager will work closely with the broker to present a proposal to the board to show that:

The cover required has been sought in all insurance markets.

That if the cover is spread over different markets the company is adequately protected, particularly when changes from current cover are proposed.

That the insurers are reliable.

That the prices offered are competitive.

The methods of payments available, i.e. deferred, quarterly, etc. and the cost advantages through discounts and cash flows associated with them.

If the company does not have a policy in regard to the cover they require, the BOD should also be fully briefed on the areas not covered and the potential for financial loss through not arranging such insurance. Similarly they should be aware of any unusual warranties contained in the policies.

Arranging insurance is very much a market place activity with rumours of cheap insurance and reliable and unreliable underwriters and brokers. Like all business activities, trust backed by personal experience with the individuals or organisations involved, plays a large part, although the insurance manager may have to defend his choice of broker and thus underwriters and P & I Club from time to time.

In the case of hull and machinery insurance the acceptance and the amount of premiums are influenced by the ownership and its management record, the ship type and its value in relation to its GRT and deadweight tonnage, the area of trading, its flag and classification. In the case of P and I insurance the principal factors affecting premiums are the areas of trading, cargoes carried, flag and nationality of the crew, and, most particularly, the record of the owners and managers.

Once the insurances have been arranged the department's responsibilities in that area lie mainly in ensuring that premiums are paid on schedule and that P & I calls when made are met.

The principal activity lies in the searching, negotiating and proposing, but once this is over each year the main activity of the department is directed towards claims.

Claims

There are two important requirements in regard to claims:

To advise the underwriters or P & I Club through the broker as soon as possible after an event occurs on which a claim may be made.

To ensure that as much factual detail as possible is gathered and properly recorded.

In the case of incidents affecting the structure of the ship a classification society surveyor will be required to report on the damage and make recommendations as to the action to be taken if any. In addition the underwriters may appoint a surveyor. In the case of cargo damage the P & I Club will appoint a surveyor to report on the cargo damage.

Details of the events leading up to the incident and/or damage should be recorded in the ships log books and may be supplemented by reports.

In the case of hull and machinery damage of a serious nature the underwriters approval will be needed before repairs can take place. This will, of course, be in conjunction with the classification society surveyor's requirements and may allow the repairs to be deferred and to be re-examined at a later date, or may require immediate temporary repairs to be made permanent at a later date. Or the damage may be so serious that permanent repairs may have to be carried out before the ship can sail. Much will depend upon circumstances including the seaworthiness of the ship, the availability of repair facilities and local costs, the cost of temporary repairs with subsequent permanent repairs as against immediate or deferred permanent repairs.

In cases involving other ships or property and pollution, guarantees or bonds will probably be required and application will need to be made to the P & I Club for this. Similarly guarantees will be required for any repair work.

The emphasis is on carrying out the right procedures, as the more that is done correctly in the early stages the easier it will be to deal with the matter later. This is particularly so in the case of legal action, when the case may not be heard for two or three years after the event.

Unless there is a dispute with the underwriters over the payment on a claim, most legal action associated with ship insurance is associated with liability to persons, property, or cargo. Whenever possible such claims, if valid, are settled out of court by the insurers, but whether in or out of court, the insurance department will be required to work closely with the lawyers representing the P & I Club or underwriters. It is, therefore, necessary that the departmental staff have an understanding of the legal requirements in refuting claims.

Should the matter be brought before a court of law, then the department will be closely involved in ensuring that documentary evidence and witnesses are arranged. In this they will work closely with the ship management department and the relevant functional departments.

Claims for freight and demurrage may also be pursued through the P & I or Defence Club, who will provide legal defence facilities in claims against the shipowner. In this again, documentary evidence is vital.

It will be appreciated that not all claims are major and some, such as claims for crew medical treatment and expenses, are of a routine nature requiring little more than methodical book keeping and presentation of documents to reclaim sums already expended.

When a general average claim is contemplated it is important that the department ensures that all parties involved are properly informed and that, where applicable, bonds are deposited.

Advisory

As has already been seen, the insurance department must provide senior management with sufficient information for them to be able to make decisions on the risk policy for the company's ships, and, on an annual basis, the actual insurance for the forthcoming year.

With their personal knowledge and experience and that of others, as noted in Chapter Four, the insurance department is also in a position to advise all departments of the company on the avoidance of liabilities and risk and the steps which should be taken in the event of loss, damage or liability. Similarly they are in a position to advise and recommend to senior management, policies to reduce the cost of insurance through lower premiums.

In particularly they should ensure that not only senior, but middle management, should be aware of the financial dangers of breaching warranties, particularly those associated with the seaworthiness of the ship, e.g. in overloading, using incompetent crews, sailing in ice areas or deviating from the intended voyage.

In the case of war risks, prompt advice to the operators will be required if local conditions change. Although war risks policies will usually protect the owner for "safe" areas, underwriters can add or delete areas from their policies in the event of hostilities or anticipated hostilities. Thus although a renewal of the cover, at a much higher rate is possible, it is most important that the operators and the ship managers and senior management should know of the situation promptly in order to consider the action to be taken.

Thus their role is not only to ensure that the ships are protected against risk at the most economical cost, but they should endeavour to assist management in avoiding claims and losses. They should also develop a system such that if claims are unavoidable, at least they are minimised and the owners claims record protected as much as possible.

Other insurances

Because of their expertise in the field of risk it is inevitable in many companies that the department be involved in the insurance of the office against fire, theft, liabilities, etc. and both sea and shore staff for travel, health, and death insurances when not already covered by the P & I Club insurance. Similarly pensions are often included in their responsibilities.

Budget

Like other departments the insurance department must budget for its costs. Essentially this covers the estimates of premiums and P & I Club calls for the next financial year. This activity is not helped by the fact that premium renewal dates and club call dates do not usually link with the company's financial year. Thus a certain amount of guesstimating takes place.

Although the insurance budget has been covered in some details in *Running Costs* the problem of budgeting for losses is important enough to repeat as follows:

> "When considering insurance budgets, there arises the question of whether or not allowance should be made for "losses" and if so, whether this should be reflected in the budget of the other cost departments. Although the bulk of any loss will usually be borne by the insurer, the cost of the deductible is usually borne by the shipowner, except in a total loss situation. If it is decided that an allowance should be made for deductibles, then it also has to be decided whether they are items for the insurance budget or for the department concerned, e.g. the technical department in the case of machinery accidents.
>
> It can be argued that the deductible is a form of self insurance and should, therefore, form part of the insurance budget. However, as one never knows whether or not there will, or will not, be an accident or claim, it is considered best not to budget for such unknowns. But as with all budgets, consistency is the key word and the treatment of this matter should be a company policy matter and followed by all departments in the same way."

The staff

In many shipping companies staff of the insurance department grow into the job. They start as a clerical assistant and work their way up gathering knowledge and experience as they go. In the older maritime nations there are insitutional qualifications they can gain which without doubt will help them in their work.

The broker

As in all fields there is someone prepared to provide expert service and in the case of marine insurance this is usually a broker. However, the main difference between this expert and others is that he is paid by commission by the underwriters and not by the shipowner.

Generally speaking most marine insurance is obtained through brokers and in some markets, such as Lloyds of London, the underwriters will only deal with a broker. Thus there are times when the owner can only choose the broker but not the method of obtaining the insurance.

Like the department, the broker's work does not cease on obtaining the insurance and he will be closely involved in the pursuit of any claims.

In obtaining the best insurance and finalising claims he will have the owner's best interests in mind with a view to future business. On the other hand, he has to maintain a trustworthy relationship with the underwriters, particularly when describing ships and owners to obtain the best insurance, as to do otherwise would affect his reputation in the market.

Chapter Twelve

Owner and ship manager

"The buck stops here"

Harry S. Truman

Responsibility for the ship

One of the reasons given for the development of shipping company regulations in Chapter Four, was the desire of the shipowner to limit his liability in the event of loss or damage through the action, or lack of action, by the staff associated with the ship, i.e. sea and shore staff.

The liability of all companies and employees, particularly senior members of companies, and the right to limit that liability, varies from country to country. In the case of ships both owner and staff can also fall under the jurisdiction of countries other than the country of registry of the ship, e.g. the Master's liability in some countries for an oil spillage.

One of the first things that any owner or manager needs to establish is the identity of the person legally responsible for the company and its property. In this guidance can be taken from Lord Denning. In the case of *H. L. Bolton (Engineering) Limited* v. *Graham & Sons Limited* (1975) he said:

> "A company may in many ways be likened to a human body. It has a brain and nerve centre which controls what it does. It also has hands which hold the tools and acts in accordance with the directions from the centre. Some of the people in the company are the servants and agents who are nothing more than hands to do the work and cannot be said to represent the mind or will. Others are directors and managers who represent the directing mind and will of the company and controls what it does. The state of mind of these managers is the state of mind of the company and is treated by the law as such."

It is a difficult point in law whether the brain, mind, or alter ego of the company, lies only with the Board of Directors or in the senior or junior management of the company and may depend upon the stature of the staff and the amount of delegation.

As an "employer" the company responsibility is usually discharged by the BOD and it bears the ultimate responsibility even though experts, such as safety officers, may be used in areas where responsibility cannot be delegated. In other words, although board members may not be experts in a particular field, they should at least be familiar with that field before they delegate responsibility for it. In shipping the persons usually considered responsible for the ship are those BOD members responsible for the management of the ship.

It is on the question of delegation and how much supervision should be given to the person to whom authority has been delegated, that significant decisions have been taken in recent years by the English Court of Admiralty. As the influence of this court in the maritime world is well recognised by international convention, its decisions are of importance to all shipowners even though some may feel they are well outside, and unlikely to be within its jurisdiction.

Until the 1960s it was considered safe for an owner to delegate responsibility for the navigation of the ship to the master. But with the cases of *the Norman* (1960), *the Lady Gwendolen* (1965), *the England* (1973), *the Garden City* (1982), and *the Marion* (1983), there has been a distinct change towards emphasis on the responsibility of the owner and management in the running of a ship.

Although each of these cases was different, conclusions can be reached on the present attitude of the English Courts as follows:

That it is the owner's responisibility to ensure the safe navigation of his ship.

That it is not enough to appoint a supposedly competent Master and leave him to carry on. The owner has a duty to supervise the Master.

That the owner has a direct responsibility to take reasonable steps to ensure that the navigational equipment of the ship is properly maintained.

That the owner should establish and maintain a proper system of management which will provide checks on the performance of the Master.

That where there is recognised danger or common fault, such as in the case of excessive speed in fog, owners should take appropriate action.

That it is not enough that staff should be given instructions, regulations, and be supervised. They should "know" and understand the owner's attitude, i.e. policy, towards the safe navigation of the ship.

Now although each of the cases from which these conclusions were developed was associated with the Masters of ships and navigation, there is no reason to believe that attitudes would be any different in respect of

Chief Engineers and other aspects of safety and pollution associated with ships; e.g. in 1983 in the case of *Hasho Industries Inc.* v. *M/S St Constantine et al.*, the US Appeal Court upheld a district court ruling that the owner was liable for cargo damage through fire, because he had not exercised due diligence in seeking the cause of engine vibration which caused the fire. It also held that the shipowner failed to take reasonable steps to ensure adequate training, through periodic drills or formal instruction, to all crew members on firefighting techniques.

Of course, the basis of any system associated with the management of a ship must be associated with the Master's and Chief Engineer's authority and responsibility. Once at sea the owner is completely dependent upon them. But as has been seen, he must supervise his staff and take action if there are any indications that their performance is not what it should be. Similarly he must ensure that they are warned of any dangers or common fault in the operation of ships which they operate.

The managerial duty to supervise can be delegated to properly qualified subordinate managers, and unless there is anything to show to the contrary, the owner is protected. But if, in an action, it is shown that there is an actual fault on his own part, i.e. the brain, he may be deprived of his right to limit liability, even though the master or one of his subordinate managers, may have caused the incident over which the legal action was brought. In such cases it is for the shipowner to prove that the cause of the incident did not flow from his own fault.

Thus it is clear, that it is not enough for an owner to provide a ship which conforms to all the regulations and classification society requirements, is properly manned with men holding the right qualifications and is supplied and fitted out with all the equipment it needs for its trade. The owner must also ensure that it is managed properly, both in the ship and ashore. It is only when he does ensure this, that it can be said that the ship is seaworthy in every sense of the word.

This is, of course, the legal view. Many owners would wish to manage their ships in that way regardless of the constraints the law places upon them. From a financial point of view their supervisory and control requirements may be much more stringent. In this they should also be careful that strict financial controls are not restrictive in regard to the safety or seaworthiness of the ship. It would be no defence to say that something had not been repaired or renewed because the budget had already been exceeded.

The owner

The owner can be an individual, owing all the stock of the company while being at the same time chairman, managing, and sole director. In such a case he shoulders the full responsibility of ownership.

Alternatively the ownership may be held by the shareholders of a company who appoint a BOD to manage the company for them. The directors may be salaried executives appointed to the Board who are also involved in the day to day running of the company as well as the long term management. Alternatively they may be non-executive directors appointed to the Board who receive fees for their services and are only involved in decisions at BOD level.

The chairman and managing director may be one and the same, or separate persons. The roles of each of them and of the other directors, or presidents and vice presidents where these are appointed, will vary from company to company. However it may be assumed that those directly involved with the management will be considered to be the alter ego/brain of the company, and therefore responsible in the eyes of the law for its mind and direction. For the remainder of this chapter the owner will be referred to as if he or she is one person.

The "owner" can sub-contract the management of his ship to a professional management company. In so doing he contractually passes the responsibility to the managing company who then take on the responsibility of "owner" as considered earlier. Such a company will also have a guiding mind or alter ego and in such respect can be considered to be the owner. The "owner" for management purposes can therefore be considered to be that person, or persons, who are legally responsible for the management of the ship.

The owner's job

It is rarely that any owner is advised that at a particular date in the future he will be given a number of ships to manage and sufficient funds to plan and establish the organisation and systems to support those ships efficiently. All too often fleets are either growing or contracting unexpectedly, and staff and their roles have to be changed to suit the circumstances. But policies and systems, in general, should not be affected by such changes. Thus the sooner these can be established and a firm management foundation created the better. The policies will probably already exist in the mind of the owner as a result of his previous shipping and managerial experience and should not take long to formulate in the areas referred to in Chapters Two and Three. But advice and comment should be sought on the policies from colleagues and senior management.

Again, the organisation and systems will probably already exist in the experience of the senior staff, and only need to be fitted to the company's particular needs. Alternatively both can be designed for the company by an outside organisation.

It is in the choice of staff to support the organisation and systems, that the owner will have to give considerable thought and attention. And although much effort will be required in this respect in the early stages of the development of the company, it should be borne in mind that staff and organisation are usually a recurring matter as people and company need change.

But firstly the owner needs to be clear in his own mind of the way in which he wants, and can run the company, e.g. he may wish to run it autocratically, receiving all information and making all decisions, but the number of ships in his fleet may preclude him from doing this effectively.

Thus he has to decide how much and to whom he should delegate responsibility. At the same time he must ensure that the systems established pass vital information to him promptly. He must know if something important is not as it should be. In this the position descriptions and authorities described in Chapter Three, combined with the control systems described in Chapter Five, will assist if not completely assure that he knows if anomalies occur. Through them his managers should know everything that is happening and should select the information he should know.

But, human nature being what it is, no matter how tight the controls and efficient the report systems, mistakes will sometimes occur and be overlooked or hidden, for a variety of reasons. The owners best protection against this is a good management climate, both in ships and ashore, by the choice and motivation of staff such that they regulate themselves, and supervision is minimal.

The role of the owner will depend upon the size of the company, and the amount of time he spends on long term and short term matters should reflect this. That is, the larger the company the more he should have responsible managers to whom he can delegate the short term, day to day management of the ships. The smaller the company the more he must be involved himself. If he is also involved in the total company operations, he must still fulfil his responsibilities to the shareholders to run a profitable company with seaworthy ships.

The owner's background

The owner may have been employed in the shipping industry from an early age and gathered his experience and training as he progressed up the promotion ladder. Ideally he should have spent time in a number of departments thus broadening his experience. Alternatively he may have commenced his career at sea and then come ashore and advanced to senior management, again through a number of departments, probably as a superintendent and perhaps as a crew manager.

As has been seen at the beginning of the chapter, he does not have to be an expert in any particular area of shipping, but he must have sufficient knowledge of the vital areas such that he is able to ensure that they are properly covered, and that his ultimate responsibility and the interests of the company are protected.

The ship manager

As mentioned in Chapter Three, there can be a number of different titles for the person ashore responsible for the management of a ship, i.e. the line or executive connecting link between the shipmaster and the owner.

Like the owner, the role of the ship manager will depend upon the number of ships in his care. It can range from the large role of Head of Fleets mentioned in the 1970s matrix organisation, to the Ship Group Manager or ship manager looking after two or three ships.

In the former he has under his wing the line and functional departments and may formulate policy within his group, such that he may be considered the "mind" of the company in ship management matters. In the latter he is more intimately involved with the ships in his care and is, essentially, only involved in the day to day and short term management.

In reality the difference between the fleet and ship managers positions is one of magnitude and as the fleet managers position is really more akin to that of the owner, it is the ship managers position which will now be considered in more detail.

Like all managers, he has to lead, co-ordinate, advise and support those who report to him. At the same time he must work closely and co-operate with others outside his department, such as the operators and the service departments and officials of other organisations such as classification surveyors and union representatives. Additionally he will be accountable to a senior member of the company or directly to the board of directors.

His responsibilities will include ensuring that the ships he manages are:

Manned with an optimum crew, i.e. the numbers, skills and qualifications are compatible with the technical, operational, and economic requirements of the ships.

Technically maintained and operated to the owner's standards and policy and in conformity with statutory and classification society requirements and insurance warranties.

Supplied with equipment at safe and economic levels in order that the programme of the ship is not interrupted.

Available as required by the operators such that out of service periods for repairs and maintenance and crew changes, are arranged in the most economical manner, having regard to the overall requirements of the owner.

Fully supported at all times, particularly in emergencies, with adequate information, advice guidance and services as required.

To his senior management he must provide:

Yearly financial (budget) and maintenance plans based on the advice he is given of the company's long term plans.

Regular reports on progress related to those plans with explanations of variances and forecasts of anticipated changes in the results at the year end.

Immediate advice of emergencies, and deviations from company policy, misbehaviour of senior staff and any other irregularity, which if not corrected or dealt with, could affect the owners legal and insurance position.

Advice and recommendations on company policy and on any other matter in which his expertise and experience may be required.

The ship manager's job

Essentially the ship manager's work should start with objectives and plans to achieve those objectives. The way in which those plans are formulated and presented to senior management for approval have been covered in *Running Costs*. For that reason, it is enough to say that plans and the cost of those plans are based on the long and short term plans for the ship and are constructed on knowledge of the ship's requirements and associated costs, such as crew, supplies, spare gear, and insurance. The plans are usually prepared by those responsible for the particular areas, for instance the crew department, technical department, etc. and submitted to the ship manager for consideration and co-ordination before completion of the overall work plans and ship budgets.

Having identified his objectives, made his plans and had them accepted by senior management, the ship manager must then "manage" and he does this through the controls referred to in Chapter Five. These should be available to him in four ways:

Firstly; the "systems" of the company will provide him with regular reports on expenditure, work progress, and operational efficiency. From these he will be able to make a numerical assessment of the performance of the ship by comparing the data with the plans.

Secondly; the "communications" between ship and shore and to and from other organisations will indicate to him whether or not the ship and staff are performing as required. Thus he or his deputy should see all communications relating to the ships and should initial, and date, them. They should also comment or take action on them as required.

Thirdly; through supervision: He will do this by arranging for a competent member of the shore staff, or an outside expert, to visit the ships and supplement these visits by visiting them from time to time himself. The purpose of the visits is two-fold: to ensure that matters are as reported through the systems, and to meet with the sea staff and listen to their points of view and receive verbal operational reports.

The frequency of such visits and the way the supervision is carried out, will depend to a large degree on the calibre of the senior sea staff and their service with the company as discussed in Chapter Three.

Nevertheless, supervision is essential and the fact that a senior officer has an excellent record and long service with the owner should not prevent systematic checks being made. The best of people "slack back" at times and the need for checks should be appreciated by competent sea staff. A record of the visits should be kept and should include details of any irregularities and action taken.

Fourthly; the manager should control through meetings both ashore and with the senior sea staff in the ship or in the office as convenient.

A meeting is more formal than a visit. Its value lies in that it is usually held aside from the routine work, i.e. it takes key staff away from their routine work for a short, but effective time, so that they can concentrate their thoughts without interruption. Ideally there should always be an agenda and minutes of the meeting kept and distributed to all those attending. Whenever action is required this should be clearly shown in the minutes with the name of the person expected to take the action.

Meetings should allow discussion on controversial matters to be raised and either decisions taken, or collective opinions obtained, for actions elsewhere. They should never be long. As a broad guideline, the meetings between ship and shore staff should take place every six months and amongst the shore staff once a month. Care needs to be taken that meetings are not postponed indefinitely. Those who never have time for meetings should be viewed with caution, i.e. if an executive cannot give an hour once a month to discuss matters of common interest with his colleagues, something is wrong.

The ship manager's job will also involve him in decision making and although decisions are not always easy, it is true that "good decisions are made on good information". Thus if the controls referred to have been arranged properly the ship manager will be in a better position to make the best decisions required of him.

As explained in Chapter Three, there is a difference in the role of the ship manager of a ship group in a centralised organisation and one organised on the SDC concept, although the differences are less than they would at first appear. The role of the former is as described above, while in the latter it leans more towards co-ordination and support, although the ship manager still retains an executive or line management role.

The essential difference lies with the sea staff, through their greater involvement and accountability. But apart from this, there is very little difference, as in both types of organisation co-ordination and support is required from the shore. With a good management–staff relationship "authority" is not conspicuous. But as was shown at the beginning of the chapter, supervision is always necessary for the protection of the owner and to satisfy his need to know that his ships are being managed properly and that all is well with them.

The ship manager's background

As has been seen, it is essential that anyone responsible for a ship must have a full knowledge of ships in general and an intimate knowledge of the particular ship he manages. Even though he may not have a full technical knowledge of every aspect of the ship, he will need sufficient understanding of the ship to be able to understand specific technical matters during discussions with experts. He must also have a good understanding of the laws affecting ships and owners and have developed an understanding of accountancy procedures, as "costs" form such a large part of his job. He must also have a full awareness of the commercial factors involved in operating a ship and have the ability to work closely with the operators.

To gain such experience, he will probably have spent most of his career in the marine industry and may have been at sea as Master or Chief Engineer before coming ashore. Alternatively he may have held a managerial position in the ship building or ship repair industries. He may have spent time as a technical superintendent and/or time in the crew department. He will also have been closely involved in the preparation of plans and budgets.

Chapter Thirteen

Master and chief engineer

Until the 1960s the very idea of linking the Chief Engineer with the Master in any form of joint management of the ship, would have been received with astonishment and probably hostility on the part of anyone from the deck department. All this despite the fact that the pay of the Chief Engineer was probably only marginally less than that of the Master, they both had the same number of stripes on their uniform, although for some curious reason only the Master had gold leaves on the peak of his cap, and the level of knowledge required for their certificates of competency was and is comparable. Neither could the ship sail unless both were on board holding those certificates.

The increased technology in ships involving all departments, the trend towards committee management on board, and the delegation of all maintenance responsibility to the Chief Engineer, drew him outside the engine room. In so doing it changed the popular idea of the deck staff that he was a convergent thinker, i.e. thinking only of his machinery, while the Master was a divergent thinker as a result of his need to be thinking constantly of the many different aspects of running the ship.

Because of the Chief Engineer's broader role and new working relationship with the Master in many ships, these two have been coupled in this chapter. However, once again it should be noted that there are still many ships sailing the seven seas, where the traditional relationships between Master and Chief still exists, i.e. the one in full charge of the ship while the other is expected to keep the machinery operative, but little more.

To consider them separately at first.

The shipmaster

H. Holman in that well-known book *A Handy Book for Shipowners and Masters*, which can be found on the book shelves of many owners, agents and captains, writes on the Master's authority and responsibilities as follows:

"To secure a high degree of success as a shipmaster a man must possess a rare combination of qualities. He must be physically sound and strong with a personality capable of commanding the necessary degree of confidence and willingness to obey on the part of his subordinates. He must be morally strong seeing that failure in an emergency so often results in disastrous consequences. No amount of experience will compensate for lack of nerve and will power and self restraint.

"Upon the skill, the honesty, the wise discretion of a master, the shipowner relies for the carrying out of business transactions in which the property at risk is frequently of huge value. The master is charged with the safety of the ship and cargo; in his hands are the lives of passengers and crew. His position demands the exercise of all reasonable care and skill in navigation, of at least ordinary care and ability in the transaction of business connected with his ship, and the constant use of patience and consideration in his dealings with those under his command or entrusted to his care."

Although this was written some years ago, its concise definition of the person required to fill the role remains the same. For having appointed the person, he is "in charge" legally and the only order he cannot disobey in the interests of the safety of the ship, its passengers, crew, and cargo, is that of dismissal. No one can make him sail or take cargo if he does not think it is safe to do so. Neither can anyone specify the precise courses he should steer or his speed to his next port, or make him sign documents if he does not think it right to do so. Of course, in all such matters, he must behave professionally and may be called upon to justify his actions, as in refusing to take cargo or sign clean mate's receipts.

His authority is large in theory but has been greatly curtailed in recent years through the ease of world wide communications referred to in Chapter Five.

Thought has been given to changing his title to "Ship Manager" or making the Chief Engineer the senior person or man in charge and perhaps relegating the Master to the old sailing ship rank of "Sailing Master". But to do so would mean changing the marine laws of many countries and the arguments are by no means strong enough yet, if they ever will be, for changing the status of the Master in the foreseeable future. For someone has to be legally in charge of the ship and although agencies have taken over much of the responsibility of the Master, particularly in regard to documentation, he is still very much needed in this role.

Outside the actual management of the ship, owners vary considerably in the involvement they require of the Master. Many nowadays only seem to require technical operators, i.e. people capable of taking the ship from port to port, carrying out all the ship management functions referred to in Chapter Two, and loading, carrying, and discharging the cargo safely.

Sometimes even the loading and discharging is virtually out of their hands, as it is dealt with by experts ashore arranging the stow with computers, be it a variety of liquid chemicals or containers.

Others still require the Master to have the old style commercial knowledge and expertise. While in some trades with first class agencies this may no longer be necessary, there are other trades in which this knowledge can be of considerable value to the shipowner. This is particularly so in the jungle of mate's receipts, bills of lading and acceptance and delivery of cargo.

For many Masters the scope of his job is in direct proportion to the involvement of ship and staff in the new systems of management referred to in Chapter Three.

At the one extreme he will be in charge of a traditional ship organisation into which some of the new techniques may have been introduced by the shore management. If they have, he will probably be required to chair management and safety committees, which will be referred to in the next chapter and also ensure that the shore management are provided with the control information they require. He will also be much more involved in personnel and welfare work than in the past.

At the other extreme, as Master of a ship in a simulated decentralised organisation the scope of his role will be much larger with greater emphasis on co-ordination and team leadership than before.

He will still chair committees as described above, but he and the Chief Engineer will be accountable for the achievement of plans and budgets, which they, supported by their staff, have been involved in constructing. They will also have authority to take the necessary actions to control the work and costs to achieve their targets. This may well be within prescribed limits, but is nevertheless a distinct change from his previous responsibilities and authority and relationship with the Chief Engineer.

In the same way there is a distinct change in his relationship with the shore staff. The tendency is more towards a "works closely with" than a "works under" relationship.

Before looking at other aspects of his job, which apply equally to the Chief Engineer, it is well to consider the other's job.

The chief engineer

His position has never been seen in the same legal light as that of the Master, and still, curiously, remains that way in some respects; as for example, in the case of oil pollution resulting from pumping out dirty ballast. In this it is the Master who is liable to be prosecuted, although it may well be caused by an engineer reporting directly to the Chief Engineer.

But he is responsible for the machinery in the ship and can refuse to operate the machinery if he feels it is not safe to do so. Like the Master, he can be prosecuted for not carrying out his duties professionally and can lose his qualifications if the matter is serious enough. Thus the responsibility has always existed, as has the authority to say no, if not always to say yes, as in matter of expenditure on repairs and modifications.

The changes referred to at the beginning of the chapter have undoubtedly broadened the Chief Engineer's role, although there are still ships where he is nothing more or less than a marine engineer, whose role is to keep the machinery operating efficiently.

At the other extreme his job is much larger. In the SDC ship he is fully accountable for all maintenance, with perhaps some limitations such as dry docking, modifications, and major repairs. But otherwise it is upon his plans and costs that the budgets are constructed, although he may have been helped by the shore staff with advice and information.

He will also be involved with the Master, in the total budget of the ship, because of the supplies needed to maintain the ship and provide sustenance for the crew who carry out the work. At the same time he will have an interest in the crew costs, because of the inter-relationship between the costs of the crew carrying out maintenance and the alternatives of riding or flying crews and, or, shore labour.

Again, because of his senior position on the management committee of the ship, he will be closely aware of the operator's requirements and will be doing his utmost to comply with them. He may not have too much time for the social committee work, but will be an important member of the safety committee.

These are the essential managerial factors in the jobs of Master and Chief Engineer. The Master may still be viewed as a "line" manager and the Chief Engineer a "functional" manager, but today the difference is less distinct. There are also three common features of their jobs which were not seen to be necessary in the early days of steam. There are leadership, control, and training.

The leaders

Leadership: One of the interesting facts about officers of merchant ships compared with naval and military officers is that they are not, generally, taught the art of leadership. They generally have to pick it up as they go, although in recent years there have been courses available in the older maritime nations on the subject of personnel relations and man management in ships.

Today, every leader should take account of the persons he leads. He should be sensitive to the needs, hopes, and fears of his staff. Leaders in ships should not only make themselves available to counsel staff about their work, but must also make themselves approachable on personal matters if necessary. It must be borne in mind that at sea there is no doctor, priest or confessor, and only rarely a wife to discuss matters with.

They must be able to communicate as described in Chapter Five by listening to what their staff say. Similarly they must be able to report effectively and make positive proposals to senior management in regard to policy and the ship.

They must also be able to motivate people by their enthusiasm and interest in their work, by their own managerial and technical expertise and by involving staff in the decision and work processes.

As will be seen in the next chapter, the Master and Chief Engineer should delegate as much responsibility and authority as possible without losing overall control. They should discuss work plans with staff, thereby recognising their skills and work experience too, and at the same time benefiting themselves from those communications.

Control: Of course their roles have always included an organising factor, and planning and control though perhaps informal, was a part of this. However, today the control aspect of their jobs is much more dynamic. They must not only "control" the safety of the ship and the protection of the environment, but must also control the maintenance and costs of the ship, aided by systems. In this their degree of involvement will vary with the management organisation of the company, but in any case they should understand the control needs, whether required by themselves or by others. In this they must accept that they too will be "controlled" through supervision.

Training: There is an old saying that as a manager you always leave something with your staff, such as some method of working, some piece of advice, whether good or bad. Thus Master and Chief need to bear this in mind in the examples they set. But there is also a more formal side to training, particularly in the technical aspects of the work, be it navigation or engine operation and maintenance. The leaders should never forget that part of their role is to teach. Much knowledge is undoubtedly acquired during training ashore and at colleges, but on-the-job training is equally important. This applies particularly to safety, and both leaders should do their utmost to see that their staff are aware of the dangers inherent in ships, and practised in how to deal with them.

In this they are aided today with films and video tapes which can be played and discussed during the voyage. Thereafter they should ensure that this education is put to practical use.

Working relationships

As has been seen, the management of a ship today is much more a joint leadership at the very top and a team leadership below. For this reason it is vital that the Master and Chief Engineer should be compatible. Both should be so aware of the dangers of incompatibility, such that they should be capable of advising the management ashore of an unsuitable relationship situation. At the same time, the shore management should be careful of the men they put together and even then keep them discreetly under observation.

The relationship between these two officers and the shore organisation will vary with the policy or attitude of the senior management and the organisation itself. Regardless of organisation, some owners include the two senior sea staff as part of the total, where in others it is as if there are two separate organisations, i.e. Ship and Store.

SDC when practised as described in Chapter Three makes this impossible, but does demand that the authority as well as the responsibility of the two are clearly defined. This would also be of assistance in other ship–shore organisations but unfortunately is rarely done. Perhaps because the shore staff themselves are also unsure of their authority.

There are other relationships too: when in foreign countries the Master and Chief Engineer are vitually representatives of the shipowner and work closely with the owner's and operator's agents, government and port officials, cargo shippers and receivers, classification society, and insurance surveyors. The number of people and nationalities with whom they officially come into contact is considerable, and this requires special skills on the part of them both, particularly after an arduous sea passage.

Considering their considerable responsibilities and the fact that they are, of necessity, out of physical reach for long periods of time, means that both Master and Chief Engineer be special people. For this reason the shipowner should take particular care who he chooses to hold these positions. Promotion to them should be by careful selection and never by seniority from the junior positions alone.

The position descriptions contained in Appendix Two give some examples of the scope of their work and responsibilities.

Chapter Fourteen

Manning and motivation

As mentioned in Chapter One, there have been a number of significant developments associated with the manning of ships since the 1960s. Principally these have been:

The design of ships.
The numbers and skills of people employed in ships.
The systems of working in ships.
The ship board organisation.
Training.
Harmony and motivation.

The design

Today it is easy to say that ships of the 1940–50s were overmanned, but this is only partially true. If one compares a general cargo ship built in 1950 with one built in 1980 it would be seen quickly that the 1950s ship was very much more "labour intensive" than the ship built in 1980. A prime example of this is the hatches and beams. The earlier ship had literally dozens of wooden hatch boards which had to be manhandled until the waterside workers in Australia insisted that they had to be fixed together into "slabs" which could be handled by the cargo derricks or cranes. Each hatch beam had to be guided into position by crew members or stevedores. But apart from the manpower involved, the time spent in opening and closing the hatches reduced the amount of time available for cargo operations considerably. And this was "operational" work which was quite apart from the maintenance work involved in maintaining the hatches and tarpaulins. Today, little more is required than the touch of a button and the hatches at various deck levels slide quietly open. The whole operation carried out in minutes by one man, whereas about eight would have been required in the past, taking at least twenty minutes at each hatch.

Crude oil washing (COW) is yet another example of equipment which has reduced the work of sea staff. Although prompted by the desire to empty ship's oil tanks efficiently and avoid polluting the sea, COW has also had the effect of reducing the man hours involved in cleaning the tanks of oil tankers.

Design changes have come about in a number of ways. Some, like the hatches, probably through market demands because of the time involved in so many shipboard cargo operations. Others because the builders and owners were seeking ships both cheaper to build and easier to maintain. For example, the incorporation of materials such as laminated plastic bulkhead coverings in accommodation, reduced the manpower costs in building through less painting and later in minimal upkeep during the ship's life.

Some builders have now designed ships for operation with very small crews, to suit what they consider to be shipowners requirements for the future. To do this they have had to study every aspect of the type of ship envisaged as follows:

Operations: Berthing and unberthing, watchkeeping at sea and in port, cargo work, ballasting and deballasting and bunkering.
Maintenance: Scheduled maintenance, i.e. planned maintenance with special attention to unscheduled or breakdown maintenance.
Support services: Catering and cleaning, including storing the ship.

They have then endeavoured to produce an optimum design to minimise the number of staff required for all the above activities. In this they have made maximum use of the most advanced automated equipment available, with related electronic and computer control, fault finding, correction and alarm techniques.

They have paid particular attention to minimising and, whenever possible, eliminating the causes of hazards such as fire and other common accidents. They have also considered the most common causes of breakdown and in addition to endeavouring to eliminate them, they have designed alternatives such that the ship can be brought safely to port if repairs cannot be effected.

In other words they have not only endeavoured to design an efficient and safe ship, but also a reliable one, recognising the relationship between these factors in the manpower requirements.

Their approach to seeking minimum crews was similar to those used by work study practioners for existing ships, which are as follows:

Numbers and skills

The first thing that should be done in such an exercise is to assess the work load of the ship. In the case of maintenance this is done by listing the tasks which can be carried out without shore assistance and the work involved. This is not as difficult as it sounds, as many pieces of equipment in ships have already been studied, so that the frequency of maintenance and the man hours of skills required are already on record. This data can be used, provided that the equipment is sited to suit easy maintenance and not in such an inaccessible place that it creates more work. It also depends on the ship having the right equipment and tools to carry out the maintenance.

The operational needs of the ship are assessed on an assumed number of voyages and cargo operations in a given period. The support services activities are based on experience and observations, being largely of a routine nature.

Having determined the work load, the next step is to decide the policy on who should do it. As far as the maintenance is concerned, this policy will depend on a number of factors, principally the trade in which the ship is engaged and the speed of ship turn around. With a high cost crew and low cost shore repair facilities, it may well be that a large proportion of the maintenance will be done by shore labour. In ships such as ferries, the turn around may be so fast that the ship's staff could not cope with the maintenance, and again shore labour may be used, or the ship taken out of service in the low season for overall maintenance. Despite the feasibility of sea staff carrying out much of the maintenance on general traders, some owners prefer this mode of running their ships, for their own economic or seasonal reasons. They may even have an interest in the repair yard of the home port.

Alternatively they may prefer to put "riding crews" or special maintenance gangs on board on certain sea passages to carry out specific maintenance. Or they may put them on board to supplement the ship's staff if the planned maintenance falls behind schedule.

The most popular mode for world wide traders still seems to be for the crew to carry out as much maintenance as possible, provided of course, that they have the necessary skills and equipment. To give the maintenance into unskilled hands can be expensive and dangerous.

As far as the operations are concerned, the decision on who carries out the tasks depends very much on the trade in which the ship is engaged and the ship's equipment. Shore workers can assist with many labour intensive operations such as hold cleaning. There have been ideas of putting mooring gangs on board ships to assist at the peak demand at berthing and unberthing. But in general, flexibility of the ship's staff together with improved equipment design seems to have overcome this. For example, in mooring operations using self tensioning and remote control mooring winches.

The difficult load to assess is that of the unexpected, such as emergencies and breakdowns. Hopefully, in ships designed as described and manned with small but highly trained staff there should be less risk of the unexpected. Accident and fire conscious staff help to prevent hazards, and automated equipment reduces the risk of operational errors. Similarly improved design and systematic monitoring and maintenance techniques reduce the changes of breakdowns.

The final factor in the manning formula is the skills requirement. It is not enough to just calculate how many man hours are required to maintain and operate the ship and decide whether they should be done by the crew or others. It is necessary to know how many man hours of which type or skill are required, and then assess the composition of the crew.

In this the ship manager has to bear in mind the restraints of government regulations and unions, in order that any change from the traditional manning of ships of a particular type and size, may be negotiated and approved. In countries with strict government departments and, or, strong unions, such negotiations are usually helped if it can be shown that the work of the ship and its manning requirements have been scientifically and practically studied.

The actual numbers have a magic ring to them. First the 30 crew barrier was broken for a 35,000 tons ships, then the twenty crew barrier was broken and even as a 17 man crew is achieved a 12 man crew is found to be theoretically possible. This is subject of course to considerations of safety, particularly that of adequate rest for the staff involved in a multitude of tasks.

It is noteworthy that as a result of the reduction in the labour intensive work in ships, the ratio of ratings to officers has changed from a crew of the 1940s of 2:1 to 1:1 with every indication that this will swing further, to a situation where there will be only one rating to every two qualified officers or technicians. This does not mean that the ratings are themselves unskilled, but that there is very much less work of a purely manual nature in a modern labour-efficient ship.

The system of working

Although there are many variations, it can be said that there are three types of systems of work in ships. The traditional, the General Purpose (or GP), and Inter-Departmental Flexibility (or IDF).

The traditional is that of the three department type existing in ships prior to the innovations of the 1960s and later. There are strict demarcation lines between the departments with each only interested in his own tasks, and only involved with the others while going in and out of port and in emergencies.

With GP there is a common work force of ratings within some semi-specialists, such as helmsmen and motormen. The officers are still in departments, but the seniors make joint decisions on how the ratings should be utilised. For this broader role the ratings must be trained to carry out work both on deck and in the engine room, up to a certain level. The flexibility of such a work force has a positive advantage over the traditional system of crews with strict demarcation lines.

The IDF system allows the ratings of a traditional crew to work in other departments at work compatible with their normal departmental work, for an agreed number of hours each month. Thus a seaman can paint in the engine room and rig tackle and stages for overhead work, because this is no different to his usual work. In the same way an oiler can oil the working parts of equipment on deck and can operate winches and other mechanical equipment. Stewards can also sweep decks. Again there is very little change in the officer's traditional role, except occasionally, for mates assisting in major rigging work in the engine room and engineers assisting with maintenance work on deck. Like GP, the senior officers are required to decide when and where the flexibility is to be utilised.

Since GP and IDF were developed in the early sixties, there have been a number of further developments towards greater flexibility, particularly amongst the Scandinavian and North European shipowners. In some the officers have moved towards dual roles, i.e. their official role and a lesser assisting role elsewhere, either in their own department or in another. This will be considered later in the chapter.

It is noteworthy that for years many thought that the radio operator with time on his hands in port was very under-utilised, but the system generally precluded him doing anything else unless he volunteered. Today the increased amount of electronic equipment for which he is responsible, including radar, and short turns around in port, has nullified these comments in large ships. In small ships the reliability of radio telephone communications referred to in Chapter Five has made the radio operator obsolete, even on ocean passages.

However, in the main, the roles of the officers have changed little except in the areas of responsibility and accountability, as described in Chapter Three and this will now be considered:

The shipboard organisation

The shipboard organisation, or hierarchy as it sometimes known, of the 1940s looked something like this:

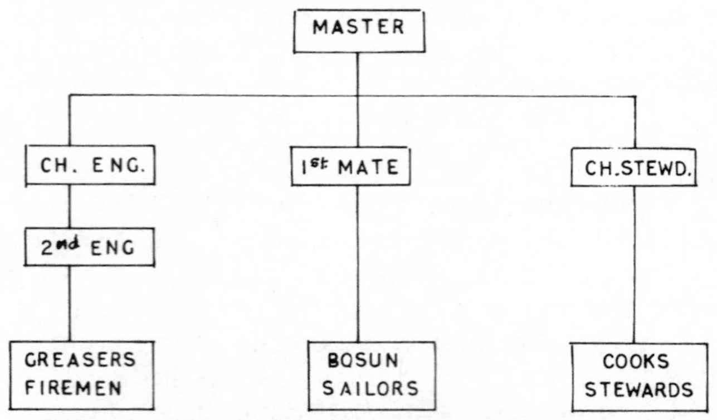

Diagram 11. The Shipboard Hierarchy of the 1940s.

As a result of the various studies of ships and shipping company organisations, a certain amount of rationalisation took place in many of the companies seeking greater efficiency, of which the most significant was the decision to make the Chief Engineer responsible for all maintenance in the ship. (See also Chapter Thirteen.) Another change was the rationalisation of non-victualling supplies, so that the ordering and control of common stocks was placed in charge of one person. Thus the mate ordered the paint and cleaning materials for all departments instead of each ordering his own as in the past. Similarly the Chief Engineer ordered all oils and greases and maintenance equipment.

But these were essentially changes of role and although the traditionalists were concerned at what they saw as a reduction in the mate's role, he was still involved in the maintenance and had many other tasks to perform. The greatest change was in the hierarchy or "the top four", and this was a direct result of the demands of the new systems of work which required collective decisions on the optimum use of the work force on a day to day basis.

To achieve this a team or committee leadership was necessary and this had the effect of flattening the organisation structure pyramid into a trapezium shape, as had happened in many other organisations. The organisation thus looked more like this:

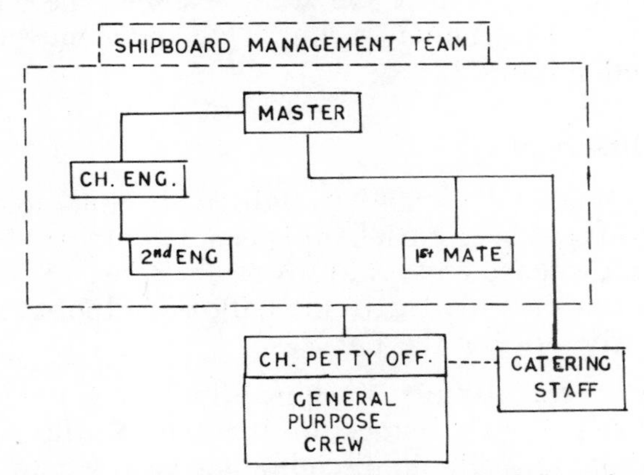

Diagram 12. The New Shape of the Shipboard Organisation

Many of the sea senior staff had to be taught the new techniques, and to some brought up in the traditional ways, the new concepts of equality at the top were difficult to accept.

Again it should be stressed that one does not have to have a GP or IDF type or organisation to have a team or committee style leadership. Many techniques necessitated by those new systems, can be applied to a traditionally manned ship. They are just more effective when the work force is flexible.

In ships which have been studied and whose crew needs have been assessed, it can be expected that the number and mix of specialists, semi-specialists and non-specialists will vary with the ship's particular needs, and the owner's technical policy, as discussed earlier in the chapter and in Chapter Nine. But the breadth of their role will depend upon the type of involvement the owner wants from them, i.e. whether he just wants operators, or a crew involved commercially through involvement in every aspect of ship management, such as planning, budgeting, controlling costs, etc. The advantages of this involvement will be considered in the last section of this chapter.

Training

The organisation for training crews has been mentioned in Chapter Eight, therefore it is enough to say here that the organisation of crews to suit new systems of work usually requires key members of the crew to be specially trained, not only in the new systems, but to accept the "change" from the old systems.

In the extreme case of an SDC organised shipping company and ship, one cannot expect the sea staff, particularly the seniors, to pick up skills for which most managers require some training. In this there are advocates for management science to be included in the syllabus of certificates of competency, but it may well be some time before this is considered as essential to running a ship. Thus the shipowner should arrange such training himself.

Harmony and motivation

Any sailor of the times before 1940 would see a remarkable change in people in ships today. The atmosphere is generally more relaxed, protocol is less in evidence and uniforms are rarely seen. There are also fewer seafarers from the older maritime nations and more from the new nations, particularly those of the Far East.

Presumably some of the attitudes in ships prior to the 1940s stemmed from the days of sail, when strictness on the part of officers and petty officers was essential to maintain discipline in ships, where life, on and off duty, was extremely hard. To show kindness and understanding was a sign of weakness and to consider the feeling of staff was, generally, beyond comprehension.

In the older maritime nations the easing of living and working conditions in ships, education in personnel relations matters, and a change in employer–employee attitudes, all contributed to the change. For the new maritime nations, who did not have the millstone of tradition in shipboard interpersonal relationships to contend with, the way was perhaps easier from the beginning. In these, different cultures result in different working relationships between officers and crews, even when they are not all from the same country.

Some problems remain, as although the ex-colonial ratings of the old maritime nations are now almost gone, there are still many ships in which the officers and ratings come from different countries. In such ships language is a barrier, and could be dangerous if an emergency should occur. There is, of course, some communication and harmony but at a very low level and certainly not of the level required to motivate to peak productivity.

For one can have a harmonious ship without necessarily having a productive ship in the managerial sense of the word: i.e. safe, performing to requirements, maintained to schedule and within "budgets". One can also have productivity without harmony. It has been found that a tough competent management, both ashore and in ships, with strict control systems for safety, performance, maintenance and budget, can produce the required results. But at a price; as it is unlikely that in such a working environment there would be any crew stability. Thus staff turnover would be high and recruitment costs similarly affected.

So what is sought is the motivation of sea staff such that they will want to work well and remain employees of the company, as mentioned in Chapter One. In this the shipping industry has been aided by a number of key studies in other industries, by eminent industrial and social psychologists in the USA and Europe, and by the Tavistock and BSF studies of British shipping companies in the United Kingdom. (See Chapter Three).

The subject is vast and fascinating and the studies changed thinking in many industries on what motivated people to produce more. Hertzberg, famous for his job hygiene theories, highlighted the different factors which satisfied and dissatisfied people about their work. He showed that it takes more than good conditions of service, good working environment, and security of employment to motivate people, although they will be dissatisfied if they don't have them. Satisfaction, i.e. job satisfaction, comes from a number of human "needs" of which the prime are:

The need to have role and status clearly defined and to have recognition of them from associates.
The need for career advancement opportunities.
The need for fulfilment at work.

Considering these points in relation to crews:

Dissatisfaction: Each maritime country has its own "norms" of employment for seafarers, which have developed through time or by imitation of other seafaring nations, their own industrial development, and conditions ashore in their country of origin.

Dissatisfaction will occur when the conditions of service, referred to in Chapter Eight, are less than expected, or efforts are made to deprive them of what they expect.

Satisfaction: Role and status have already been covered to some degree in the comments on job descriptions in Chapter Three. Perhaps one of the advantages of the organisation of sea staff in the past, compared with other industries, was that everyone knew his role and status and that of everyone else in the ship. They also knew the qualifications certain officers and ratings had to have before they could hold a particular position. Within the sea staff organisation they also knew their career prospects. In many respects it was a very tidy society with little political infighting, as may be seen in some large organisations ashore, presumably because it was impossible to achieve anything by such behaviour in ships.

It was only in the progression of the career ashore in the industry that difficulty was encountered, as unless one was the right man, at the right time, there was little chance of being offered the traditional position as superintendent. In many shipping companies this understanding of roles and career prospects still exists, although the availability of positions of senior officers in ships usually reflects current shipping economic conditions.

This does not mean that position descriptions are not necessary, as they help to revise the individual's concepts of his responsibilities and authorities. For shipping companies operating "new" systems of work with new staff roles, position descriptions are essential. This is not only because people are more comfortable if their job is described, but because of the need to stress the difference from the traditional ways, which tend to linger on for years unless positive steps are taken to bring about the change.

Fulfilment: The need for personal fulfilment in work is at the heart of motivating people towards peak performance. It has been mentioned in Chapter Thirteen that the enthusiasm, and dedication of the master and chief engineer in their work and interest in their staff, are important factors in motivating people. But more is required, and this is usually found through greater personal involvement of all the staff in decisions affecting their work.

This can be achieved, as in other industries and in the shore office, in a number of ways:

By making the seniors accountable for their areas of responsibility. This has already been considered for the Master and Chief Engineer in Chapter Thirteen, but by extension can include the First Mate and Second or First Assistant Engineer and Chief Steward.

If accountability, as associated with an SDC organisation, is not feasible, then as much responsibility as possible, with related authority, should be delegated to the senior sea staff. Similarly, within the ships the seniors should delegate as much as possible to the lowest possible level.

This does not mean that supervision of work at all levels should cease, but that is should be with a different emphasis, i.e. it should be directed towards counselling and discussion, rather than just checking that someone is working and how much he has done.

Involvement can also be achieved in a number of other ways, through interpersonal communications as follows:

Regular management meetings of seniors to discuss the progress of plans and seek ways of achieving future results by distribution of labour and, or, inter-departmental assistance and co-operation. These should be followed by short meetings with staff involved in organising the actual work, if they have not been co-opted to the management meeting. Their views should be sought on the best way to carry out the tasks and their estimates of the time for completion of the work sought.

Similarly meetings on safety should be arranged as regularly as possible, involving representatives of the staff. These should be aimed at highlighting areas where further training is required and identifying work sites or practices where danger exists, with suggestions for improvements.

Social meetings should also be held from time to time with staff representatives to discuss any "dissatisfactions" and to seek ways of improving the social life on board.

Staff should also be involved in meetings on operational matters such as cargo, if those are not held in conjunction with the management meetings.

The frequency, length of time, and formality of meetings should be carefully considered. There should always be an agenda to ensure that all important matters are discussed. Apart from *ad hoc* meetings for operations and emergencies, weekly meetings should be sufficient for management matters and monthly meetings for safety matters. As Parkinson says "work expands to suit the time available" and the same can be said for discussion. For this reason the time allowed for meetings should be limited whenever possible. As a rough guide, an hour should be sufficient for routine meetings.

The interest of seniors in the work being carried out by others, and their views on progress, time for completion, etc. will also give staff a sense of personal involvement in the ship.

Finally fulfilment, or job enrichment as it is sometimes called, can also be achieved through giving individuals tasks and responsibilities outside their normal role, such as the electrician or even the Master taking charge of the ship's canteen or slop chest. But in this care must be taken that in the search for enrichment, the job is not in fact overloaded. Experiments in this broader role for some staff has met with success in other industries and ships which have practised it, and there is no reason why it should not be beneficial in all ships.

As stated earlier, the only real barrier to harmony is the lack of a common language between crew members. Tradition and culture can be obstacles at first, but with persistence and perhaps some training and education, these can be removed.

As an indication of the views of eminent writers in this field, the following are three quotations taken from "Writers on Organisations".

> *"The primary functions of any organisation, whether religious, political or industrial, should be to implement the needs of man to enjoy a meaningful existence."*
>
> Frederick Herzberg

> *"The entire organisation must consist of a multiple overlapping group structure with every work group using group decision-making processes skilfully."*
>
> Rensis Likert

> *"The average human being learns, under proper conditions, not only to accept but to seek responsibility."*
>
> Douglas McGregor

Social aspects: People in the temporary ship community not only have to work together, but to live together too. In this it has been found that the factors which create harmony and motivate the sea staff in their work, play an important part. For if people are happy in their work, they will usually be happy in their off duty periods. The removal of departmental barriers has undoubtedly helped to integrate staff and this has been furthered in ships where common recreation and eating areas are possible.

People go to sea for a variety of reasons. Some because it is their best opportunity to earn relatively high wages. Others because there is still adventure and romance in the career. Some stay at sea all their lives, while others leave early or mid-way in their working career. It will doubtless continue to be like that. But regardless of their reasons, it is important to remember that ships are only as good as the people who serve in them. Apart from the legal requirements of providing a properly qualified crew, owners and managers should do their best to ensure that the crew are satisfied with their work, and not dissatisfied with their working conditions, as described earlier in the chapter.

Chapter Fifteen

Research and development (R & D)

All businesses should be continuously seeking ways to improve their performance. If they do not do so, sooner or later they will find themselves left behind by others who have found ways to maintain or improve their profitability, in a changing and competitive industrial world. Shipping is no exception to this.

Some ship owners prefer to be at the forefront of development, while others prefer to sit back and wait for others to develop new techniques, systems, products and equipment. As mentioned in Chapter Fourteen, many technical developments in ships are the direct result of research by shipbuilders and manufacturers who themselves seek to improve their product in the knowledge or anticipation of shipowners needs. These are all directed towards ultimate reductions in current running and operational costs.

But although much of the technical research and development is carried out by the builders and manufacturers at no cost to the owner, at the end he has to evaluate the advantages and disadvantages of the new products and decide whether to adopt them. More often than not this involves an extraordinary capital outlay, and although the manufacturers will usually provide a detailed cost benefit analysis spread over a number of years, the responsibility still lies with the owner.

R & D into trading is usually secretive because of the competitiveness of the industry. This does not mean that the shipowner has to carry out all the research himself, as he can use consultants expert in market research to seek the information he requires. Having decided there is a market worth entering he can "develop" his ships to suit the market, through new building or modifications to existing tonnage, again with the aid of naval architects or specialists in cargo handling and access equipment.

Today there is a trend for those selling systems and products and even those hired to carry out research, to offer to relate their fees to the savings they believe can be effected. In trading opportunities they may also offer a reduced fee for a share in the envisaged profits. Although this may reduce the hoped-for profitability for the shipowner, it does have an advantage in that it involves the experts and is thus a form of safeguard. But again, at the end, the shipowner has to evaluate the factors himself and make a decision. Even if the research and possibly the development is carried out by others, the evaluation itself is a time-consuming business, outside the normal day to day activities of running and operating ships. If there are no special staff for this purpose, it is inevitable that the existing staff will be drawn into consultation on the development. In this one turns again to the restraint of time; if the staff are finely balanced to suit the day to day tasks only, then there is bound to be a shortfall in the work they should normally do if they take on this extra work. Because of the interest in the development task, they are more likely to neglect the routine work.

R & D is like training: the owner can do nothing and wait for others to do the work for him. But if he does, he must accept that he will inevitably be left at the tail end of development in the industry.

As has been seen in this and earlier chapters, research and development covers a wide range in shipping organisation in ships and ashore, in accountability, control of costs, communications, computer and control systems and automation, cargo handling and other operational systems and everything associated with maintenance.

As with everything else in ships, it is important to stress the need for compatibility between staff, systems and equipment. It can be expensive and even a dangerous mistake to install or build equipment and systems which staff do not have the capability of operating. Thus when considering development of anything in ships, the owner must always consider whether the staff also need development to suit.

Chapter Sixteen

Operations — (in relation to ship management)

"The parts exist in contemplation of the whole."
Peter Drucker

The relationship between Operations and Ship Management functions is usually distant. This is because they are often not of the same organisation, as in the case of chartered ships, or when the owner operates his own ships but sub-contracts the management elsewhere. Even when both functions are carried out within the same organisation and in the same building, it is not unusual to find the operators acting like time charterers towards the ship managers in regard to the ships' performance and *vice versa.*

In many cases the functions are carried out in offices continents apart, while their point of common interest, the ship, is constantly on the move around the world. Often, the operators have no direct contact with the ship managers, their communications being by telephone, telex and letter. Their relationship with the ship may be similar, the only direct contact being through their agents around the world.

Despite these distances and lack of direct contact, each exists in contemplation of the other. Neither can carry out their functions effectively without the co-operation of their counterpart.

Operations itself covers a wide field, including the sale and purchase of ships, bunker supplies, time and voyage chartering, voyage costs, freight rates, conference rules, agency and legislation in regard to the carriage of goods by sea. Fur further information reference should be made to specialist books on these subjects. The purpose of this chapter is to highlight the functions of operations in relationship to ship management and to complement references to operations made throughout this book.

As has been seen, the prime function of ship management is to ensure that the ship or ships placed in their charge perform as required, within the restraints described. The prime function of operations is to seek suitable business for the owner's ships, even to the extent of chartering them out to others to operate. Alternatively, if the amount of business they find exceeds the capacity of their ships they themselves may charter in to carry the excess.

It is for this reason that the operators have a responsibility to advise the owners on the type and number of ships required for the trades in which they are engaged and on when to buy, sell or lay-up ships. The ship management involvement in this, through the provision of advice on cargoes and the ships themselves, has been considered in Chapter Twelve.

Having found the cargo and the ship to carry it the tasks of the operators *in relation to the ship* can be considered under the following headings:

The voyage schedule.
Appointment of agents and stevedores.
Loading and discharge arrangements.
Bunker arrangements.
Voyage instructions.
Costs and controls.
Claims.

The voyage schedule

Depending upon the contract of carriage for the cargo, the operators must decide the port or ports or loading and discharging, the order of call at each of the ports and, in some cases, the route to be taken. In liner shipping the loading ports may be decided on the availability of cargo and shippers requirements. In the bulk trades the loading ports may be stipulated as part of the charter party or contract of affreightment and the discharge within a range of ports.

If there are any doubts or queries about the suitability of the ports or the ship intended for the voyage, the operators should consult the Ship Manager or other expert.

Appointment of agents and stevedores

Unless the operators have their own offices at the loading and discharging ports, they will appoint agents to take care of their interests and to act for them as appropriate. At times the contract of carriage may call for the use of the shipper's or receiver's agents, although the operators may still appoint their own "protecting" agents if any conflict of interest is perceived. Similarly the operator may appoint the stevedores through their agents although, again, the terms of carriage of the cargo may dictate otherwise.

Ship Managers usually use the protecting agent if one is appointed, otherwise the operators' or charterers' agent will be used for the provision of support for their needs, unless they have a major task requiring specialist agents.

The shipmaster and crew will need to work in harmony with both the agents and stevedores if the best results are to be achieved. As we have seen, much of the responsibility for the documentation, acceptance and release of cargoes has been taken from the Master in a number of trades. Nevertheless, he is still responsible for the safe carriage of the cargo and should report directly to the operators and his owner or ship manager if he is dissatisfied or suspicious of any matter associated with the cargo. This applies particularly to the condition of the cargo and the way in which it is loaded and discharged by the stevedores. Likewise although the operators or the charterers appoint the ship's agents and stevedores, the Master and crew should work closely with them and take responsibility for their actions, depending on the circumstances of the trade.

Loading and discharge arrangements

If an owner operates a line service with his own ships, the operations department will make all the necessary loading and discharging arrangements, either directly or through their agents. But if the cargo is carried on free in and out or similar terms the arrangements are made by the charterers, shippers or receivers. In such cases the operators' responsibility lies in ensuring that the ship can reasonably reach the designated loading and discharging ports and be ready to load and discharge as required by the contract.

In both cases it is for the Master to endeavour to have the ship at the loading and discharging ports when required and to ensure that the cargo compartments are ready and cargo equipment working. Any failure in readiness or in equipment which cause delays can result in claims against the shipowner or manager.

Bunkering or fuel arrangements

The supply of bunkers or fuels is usually the responsibility of the operators of the ship. In making plans and arrangements for the supply of fuel or fuels to the ship for the forthcoming voyage the operators must bear in mind:

The ship's fuel capacities.
The fuel specifications.
The bunker ports en route and availability and prices of the fuel.
The ship's consumptions at full and economic speeds.
The cargo deadweight and loadline zones en route.
The reserves of fuel to be carried.

The Master's responsibility in this is essentially twofold: to check that the specifications of the fuels supplied are correct and, most importantly, to ensure that there are sufficient quantities of fuel on board at all times. With a deadweight cargo he must also ensure that the cargo and fuel requirements are compatible with the loadline zones through which the ship will pass.

Usually, the Master will be required to keep the operators advised of the fuel situation in the ship. He may also be required to give estimates of the fuel remaining on board and requirements at different stages of the voyage. If supplies are running low, or if he considers there is insufficient fuel for the forthcoming sea passage, or to the next bunkering port, then he must advise them of the situation.

Thus it is the operators' responsibility to ensure that the ship is supplied with sufficient fuel, but it is the Masters' responsibility to warn them of an anticipated fuel shortage. He should not sail unless there is sufficient fuel on board, including sufficient reserves for bad weather.

Voyage instructions

These are the prime communication between the operator and the ship. Although the instructions are written to the Master the ship manager should be fully aware of them, in order that he can support the ship to the full.

The instructions are usually in a letter form, although they may be issued by telex, and are supplemented as necessary throughout the voyage. They usually cover the following basic points, with additions to suit the particular trade.

The voyage: The Master will be told where to proceed and when the ship is to be available to commence loading. Thereafter he will be expected to follow the voyage schedule as described at the beginning of the chapter. This may include an itinerary if on a liner berth, and commencement and cancelling dates if on a voyage charter.

The cargo: He will be advised of the prospective cargo and of any special requirements for pre-loading, loading, carriage and discharge. He may also be instructed on the rejection of unsuitable cargo and in this respect the signing of mates receipts. He may also be asked to authorise the agents to sign Bills of Lading on his behalf.

In highly specialised trades in which the operators have a long experience, he may be issued with detailed instructions on the ports to be visited and their various customs and requirements.

The agents: He will be told of the operator's representative at each port at which the ship is expected to call and will be given full details on how to communicate with them.

Communications: He will be given full instructions on when, what, and how to communicate with the operators and their representatives. The frequency and details of such communications will vary with the trade and the operator's individual requirements.

Speed and bunkers: The operators will advise the Master of the speed at which they wish him to proceed on each sea passage and advise him when and where they intend to supply fuel to the ship.

The contract of carriage: The Master's involvement in the contract of carriage of the cargo will depend upon the type of contract. In the liner trades he may never see the Bills of Lading, whereas on a voyage charter he will either have a copy of the Charter Party, or at least be advised on the salient features on which he is expected to act, such as giving notices of arrival and of readiness to load and discharge.

Warnings: In some trades warnings may be given to masters about illegal practices, trickery and even such dangers as piracy, so that he can be on his guard against them.

These are the principal points included in most voyage instructions. As with all communications, their aim should be also to inform, guide and assist the Master. On the one hand the operators need to take care that they do not instruct the Master to do anything that would conflict with his responsibilities to all parties involved in the voyage.

For his part, the Master should do everything he can to co-operate with the operators, bearing in mind these responsibilities. As an example, the operators should not arrange cargo unsuitable for carriage in the ship concerned. If they do, the Master should refuse to accept it. Fortunately in most cases of experienced and professional operators such cases rarely arise.

The same applies to speed: the operators should never word their instructions on the required speed in such a way that the Master could feel he must maintain that speed in all circumstances. Of course, it is the operator's and the Master's job to try to avoid delays, but the Master must do so *with safety*, and the operators should always recognise this. Similarly the operators should not send a ship to a dangerous area or port where there is a risk of war or other hazards. Apart from the fact that the ship may not be insured for such risks, it is not in anyone's interests that the ship, cargo, and crew should be exposed to such dangers.

Costs and controls

Operators are no different to other managers in that they are responsible for their financial results. Unlike ship managers, the definition of accountability is more easily applied to them, as they have control over the earnings of the ship, much of the voyage costs and to some degree the running costs, as follows:

Bunkers and fuels: The operators will endeavour to arrange fuel supplies at the most economic ports. But they will also monitor carefully the consumption of fuels in the ship at sea and in port. If they are in excess of the ship's specifications they will seek to rectify the situation, as not to do so will increase their costs. In an owned ship this will usually be dealt with directly. In a chartered ship the matter will undoubtedly be raised with the owners.

Cargo handling costs: Where these are for the account of the operators they may be able to choose the stevedores and negotiate the costs. But the costs may be fixed by port regulations and the only decisions they can make regarding costs may be whether or not to work overtime at weekends, or at night.

Although the ship's staff are not usually involved in the decisions of such costs, they can affect the overall cargo handling costs by their readiness, or otherwise, to receive and discharge the cargo. Similarly, where ship's gear and equipment is involved in the cargo operations, its efficiency can directly affect the costs of the shore labour and associated costs. Thus even a delay in rigging a gangway on arrival alongside a berth can cause additional costs if the stevedores are standing by waiting to work.

For this reason, operators usually require full details of port operations, including reasons for any delays. These are often produced by the agent as a statement of fact or time sheet and signed by them and the Master. When special cargo-handling equipment is hired the Master is often required to verify its use.

Agency fees: These may be negotiable or according to a fixed scale for the port. Although the Master and ship's staff can have no effect upon these costs, the way in which they co-operate with the operator's agents will undoubtedly assist them in carrying out their tasks efficiently.

Port charges: Again many of these are fixed and outside the control of the Ship Master. Pilots have to be used according to local laws and even the number of tugs for a particular size of ship may be decreed by the port rules. However, in some ports there is flexibility in the matter of tugs and the Master can sometimes save the operators costs by querying the use of a tug or tugs, if he does not consider such assistance necessary. The number of ship movements and tugs used is usually included in the statement of fact such that the operators can monitor, if not always control, such costs.

Cost of the ship: The actual cost of running the ship is a major factor in the final calculations on the profit or loss of the voyage. Where the operator charters the ship, he has a control over the costs in his acceptance of the charter hire rate and in the fact that it is usually on a fixed cost per day. Where the operator owns the ship and the managers are part of the same organisation, fluctuations in the actual running costs compared with budget can have a direct effect upon the operator's results. Significant increases in these costs may result in the operators trying to improve their earnings, or alternatively taking the ship out of service or selling it.

For this reason, operators and ship managers in a complete ship owning organisation need to consider their relative positions carefully. There are advantages in the competitiveness of the operator–charterer type relationship, particularly in keeping the ship to its performance targets. But, with flexibility and co-ordination between the two departments, advantages can be gained which would not be possible in separate financial organisations. For example, if the operators have no work for the ship for a week or so, the time can often be used effectively in carrying out maintenance which would normally take the ship out of service when required by the operators.

Claims

Claims for shortage or damage to cargo often have to be dealt with by the operations department and although responsibility may or may not be that of the shipowner, full communication between operator, ship and ship manager is essential if claims are to be refuted or minimised. The Ship Master should therefore advise the operators immediately of any incident affecting the safety of the cargo, although the legalities of the owner's position should be protected at all times.

Conclusion

Although in some cases operators and ship managers need to protect their relative positions in regard to the safety of the ship and cargo, in the main the way in which they work together should always be for their common good, the success of the voyage.

Chapter Seventeen

How to manage ships — a conclusion

"Expenditure rises to meet income."
Parkinsons Second Law

The competition

There is no question that some shipowners and managers are in a more advantageous position than others. They can choose their country of operation, the country of registry of their ships, buy ships wherever they wish, take advantage of building subsidies and tax laws, and engage crews from a number of countries at relatively low costs.

They may even be able to run ships with a very poor record and yet still be able to obtain full insurance at competitive rates and obtain business for the ships from reputable operators. Some even have the monopoly of carriage of their country's cargo and therefore do not have to be competitive at all. All this, while others, whether they like it or not, have to run their ships to strict rules and within other restraints and have no subsidies or tax advantages, nor choice from whence they buy their ships or engage their crews.

Thus some have to strive harder to be competitive and with so many disadvantages the only way they can survive is to be more efficient than the others.

But these advantages and disadvantages aside, the ship manager's job is to make the best of what is available to him. Thus having been given ships to manage he has to decide the best way to manage them.

The decision

There are four ways one can manage a ship: safely, dangerously, efficiently and inefficiently, and as one would expect, any number of variations in between. Availability of money is often a key factor, associated with the owner's long and short term plans and policies.

Assuming the owner does want his ships run safely and as efficiently as possible in his particular circumstances, then he should make his decision on how he wants them managed in the light of the following seven prime factors:

The number of ships he manages.
The type of ships.
The age and development of the ships.
The number of years he intends keeping each ship.
The crews available to him.
The funds available.
The managerial experience available to him.

The number: Is an important factor because if only one or two are owned it may be economic to sub-contract all or part of the management functions.

The type: Different ship types often require a different type of expertise, particularly technical, so that if a ship owner is taking on a new type he will again have to consider how best to provide that expertise. If he is starting with a new fleet he will have to provide appropriate expertise for the ship types.

The age and development: Ships of the same age will differ in the amount of technical development, although in general it can be said that the younger the ship the more efficient it will be. Thus the age and stage of development will have a bearing on the crew choice, technical expertise requirement ashore, and the amount of control systems which can be installed.

The number of years of intended service: Will have a bearing on any maintenance plans and the installation of control systems. The older the ship and the shorter the time in service, the less effective will be any planned maintenance or control system.

The crew: Much will depend on whether the owner has a choice in this. If he has to use a crew from the country of registry and they are not highly trained, then it may be dangerous to put them in a sophisticated ship where they will not be able to use the equipment efficiently. Similarly he will be limited in the amount of controls he will be able to maintain effectively. On the other hand a highly trained crew would be wasted in an old, unsophisticated and labour intensive ship.

The funds available: This is often a limiting factor in any management decision. The funds available usually relate to the surplus after earnings and operational costs, and can therefore vary considerably with time, ship type and operation.

Funds will affect the choice of the crew type, if there is one and this will influence the way in which the ship can be controlled. Lack of funds can also effect the installation of control systems, including the communications systems, although from a long term point of view they would be beneficial.

The management experience available: The owner may wish to operate ships with tight controls and regulation, but if he does not have staff capable of installing and maintaining such systems, he will have to lower his target or seek the expertise elsewhere.

From this it can be seen that there are a number of choices, but any decisions on how to manage a ship, or ships, should relate to the seven factors, with emphasis on the balance between the ship, the crew, shore staff and the systems.

Once the owner has decided on the way he wants to have them managed, he should bear in mind the following general points which apply to the shipping industry in particular.

Simple and small

As has been seen throughout this book, shipping is a *detail* industry, not only in the many tasks involved, but in the amount of information essential to carrying out those tasks. It is also an industry in which *responsibility* plays a large part and this demands regulation. Although the basics of regulation are provided by Governments and some industrial institutes, there is still a need for much self regulation in the detail by the shipowner. It is also a *commercial* industry and this demands control, not only of what is done, but of what is spent. But most importantly it is a *human* industry in which the people in ships have an unusual and vital role. The way in which they work and live together has a large bearing on the success or failure of shipping companies.

As can be imagined, there is considerable scope in all these for overdoing or underdoing. For developing large organisations and systems, or employing experts beyond their required need. Or for trying to run ships on a shoestring, in the mistaken belief that ship management is only a matter of flair, or that there is no need to bother with detail or people.

From lessons learned in other industries and the shipping industry itself, some things are clear: someone has to be responsible for the ships and the basic functions must be covered. All or part can be subcontracted and this should always be considered as an alternative when organising or re-organising a company.

But the two most important factors are simplicity and size:

Whatever the organisation and its associated systems, the owner and his staff should always seek, and continue to seek, the *simple* way. Safeguards should be built in to prevent any system and organisation becoming larger than it needs to be. Staff should always question the need for anything that creates work but does not produce results.

In the same way that the philosophy "small is beautiful", has became popular, so small management units have been found to be more effective in caring for ships. Focus of attention on the ship as the prime unit has proved beneficial and has re-emphasised to ship managers, the point that the ship is the most important unit of the shipping company. Without ships there would be no shipping company. Similarly, without competent staff ships would not be run properly.

There is no one way to run a shipping company, but there are some right ways and some wrong ways. It is hoped that the good owner and manager can tell the difference.

Reference books and papers

Managing for Results, Peter F. Drucker
The Practice of Management, Peter F. Drucker
Drucker on Management, Peter F. Drucker
The Business of Management, Roger Falk
Crew Costs — The Shipowner's Dilemma, Galbraith's Shipping Advisory Services Ltd
Running & Maintenance of a Fleet of Bulk Carriers and General Cargo Carriers, T. W. Major, CEng, FIMarE, InstMarEng, 18.10.77
Optimising Ship Repair and Maintenance Costs: A Systematic Approach, J. B. Bunnis, BSc, CEng, NE Coast Institution of Engineers & Shipbuilders, 8.10.73
A New Approach to Ship's Maintenance, B. K. Batten, MSc, CEng, FIMarE, Journal of Ship Repair and Maintenance, February 1975
Handy Book for Shipowners and Masters, M. R. Holman — The London Steam Ship Owner's Mutual Insurance Association Limited
The Shipping Industry, Victor Dover
Marine Insurance Practice, R. H. Brown
The Principles of Marine Insurance, Harold A. Turner, ACII
Management Techniques, John Argenti
A Manager's Guide to Work Study, Owen Gilbert
The Directors Guide to Computing and The Directors Guide to Computers, The Institute of Directors in collaboration with the National Computing Centre
Micro computer experience in the Shipping Company, Computers, Fairplay, 18 March 1982
Texaco adopt computerised planned maintenance system, Fairplay, 26 May 1983
Information Technology for Ships, Fairplay, 25 November 1982
Up the Organisation, Robert Townsend
Condition Monitoring, Lars Noyen — Det norske Veritas, International Maritime Conference, Europort, 1975
Improving Management Performance — Management by Objectives, J. W. Humble
Writers on Organisations, D. S. Pugh, D. J. Hickson and C. R. Hastings
The Principles and Practice of Management, E. F. L. Brech
Emerging Organisational Values in Shipping, M. H. Smith and J. Roggema, Marit Pol Mgmt 1980, Vol 7
Royal Navy Queen's Regulations
Occupation of Seamen in Japan, Marit Pol Mgmt 1980 Vol 7
Manpower Systems and the Case for Change, Nautical Review, March 1980
West German Experiments, Sea Trade, August 1981
Ship Manning, Rules & Shipowners, Sea Trade, October 1981
Sealife Research Project, Sea Trade, August 1981
Training on Board to Respond to Emergencies, R. Tooth, Fairplay, 13 January 1983

Design Considerations for Ships of the Future, M. Meek, Manning Present & Future Seminar, Hon Co MM & Naut Inst, February 1983
Shipboard Automation and its Future Potential, D. N. Loynes, Manning Present & Future Seminar, Hon Co MM & Naut Inst, February 1983
Social Theory & Shipboard Structure — Some reservations on an emerging orthodoxy, Nick Perry and Roy Wilkie, Marit Stud MGMT, July 1973
Devolution and Development of Shipboard Community in the Project of Change, P. T. Quinn, PhD, Tavistock Institute of Human Relations, May 1978
The Ship as a Temporary Human Community. Future Educational Implications, P. T. Quinn, PhD, Tavistock Institute, March 1982
Manpower Systems and the Case for Change. P. Sharpe, Nautical Review, March 1980
Ship Management, Michael Grey, Fairplay, August 1978.
Philosophy of Line and Staff and The Unit President Concept, J. Keith Lowden, American Management Association, July 1965
Corporate and Divisional Staff Relationships, J. Keith Lowden, American Management Association, July 1965
Why Line Managers Don't Listen To Their Personnel Departments, T. F. Cawsey, Personnel, January/February 1980
ITF Has a Private Face As Well As a Public One, Ake Selander, Nautical Review, March 1980
Methods of Organising the Work Load on Board and the Educational and Training Implications, D. H. Moreby, Manning Present & Future Seminar, Hon Co MM & Naut Inst, February 1983
Complete Re-think in Norway, Training & Manning, Fairplay, 6 May 1982
Smooth the Troubled Reduced Manning Waters, Norway, Fairplay, 30 June 1983

Sealife Programmes

Project 1, Report 1, October 1975. Tasks and Skills aboard ship
Project 3, Report 1, December 1975. Ship & Shore: Communications, Functions and Relationships
Design Workshops — Superstructure, J. G. D. Cain & M. R. Hatfield, June 1979
Design Workshops — Machinery Space, J. G. D. Cain, December 1979
Central Manpower Supply to the Merchant Navy, P. McCowan & M. Barry, December 1978
Productivity and Hierarchy Aboard the Deep Sea Ship, M. H. Smith, January 1979
Sealife Conference Papers, Liverpool, November 1979

Appendix One

(Chapter Four Refers)

CODE OF GOOD MANAGEMENT PRACTICE IN SAFE SHIP OPERATION

Introduction and Summary

Merchant Shipping is a specialised and technical business. Its complexity has been increased during the last ten years by the extensive new conventions developed by the International Maritime Organisation (IMO)* and the International Labour Organisation (ILO), designed to improve safety and social conditions. But regulation — including the recent emergence of port state control — can only go part of the way to achieving the objective of safe and pollution-free shipping. In the end — while the Master is clearly responsible for the direct operation of the ship — the overall responsibility lies with the shipping company.

The purpose of this Code is to provide a broad framework of good practice against which management in companies operating ships may gauge their own organisation and procedures. Its contents are drawn from the best management practices of a number of different companies represented in the International Chamber of Shipping (ICS) and the International Shipping Federation (ISF). It is intended solely for voluntary use, either as a check-list or as a framework for reviewing company methods. Parts of the Code may not, of course, be appropriate for a particular company.

SAFETY and EFFICIENCY are integral to good management. They can only be the result of structured, painstaking policy and a combination of the right skills, knowledge and experience. The direct involvement of decision-taking management in these matters is vital. The attitude of an Owner and/or senior management is reflected in company policy and thus directly in the work of all the company employees. THE INITIATIVE MUST THEREFORE COME FROM THE TOP.

Adherence to the recommendations in this Code also makes sound commercial sense. By meeting at least the basic minimum standards a company will ensure that its ships are available for trading to the maximum possible extent. Time lost — through accidents, avoidable damage, correcting deficiencies, detention, or crew unrest — means more expense and less business.

The major recommendations are that:

- While the Master and the crew have direct responsibility for the technical and safety aspects of ship operation, every company operating ships should establish a department or at least designate a person ashore, responsible for those aspects of the operation from the shore standpoint. The person(s) involved should have knowledge and experience of the basic technical aspects of the ship (e.g. structure, equipment, documentation, etc.) and of the relevant national and international regulations. The existence of such a department or person does not, however, relieve senior management of responsibility for safe and efficient operation.

- Safety and operational practice should be a regular item for discussion at management meetings at all levels. Policy on these issues should be clearly defined and made known to employees.

- Management should ensure that shore-based personnel are aware of — and provide for — the needs of the Master and shipboard personnel, in regard to the safe and clean operation of the ship.

- Management should ensure that there is a sufficient number of crew on board to operate the ship and any specialised equipment carried. The crew should be medically fit, properly trained and qualified to perform the tasks required of them.

- There should be regular and effective two-way communication (1) between shore-based and shipboard staff and (2) between management (including senior management) and employees ashore and at sea. This should cover company policy on safety and operating practice.

- Proper arrangements should be established for use in the event of an emergency involving the ship. These should seek to ensure an effective and level-headed response to the incident both by the crew on board and by the shore-based staff.

- Management should review its overall approach to the matters covered by this Code on a regular basis.

* formerly IMCO.

ICS/ISF Code of Good Management Practice in Safe Ship Operation

This Code, by its nature, covers matters which are not appropriate for regulation. It is intended as guidance for all companies operating ships and does not seek in any way to define or embrace detailed statutory requirements, national or international. It is taken for granted that such requirements have to be complied with.

The guidance can only be expressed in broad terms if it is to have widespread application. Clearly, different levels of management, whether shore-based or at sea, will require varying levels of knowledge and awareness of the items outlined. Persons responsible for particular areas should have more detailed and specialist knowledge of their specific tasks. This Code seeks to provide a framework only.

While the Code is addressed to shore-based management, it is recognised that on board the ship it is the Master (as agent of management) who has the over-riding responsibility for the safe operation of the ship. Consequently, a number of the activities recommended may well be delegated to him. It is therefore for management to appoint a Master who is fully conversant with and dedicated to the maintenance of appropriate safety standards, and to ensure that all necessary support is given to him by the shore-side organisation in the performance of his duties.

1. TECHNICAL ASPECTS OF SHIP OPERATION

1.1 Strong commitment to safe ship operation and prevention of pollution should be a paramount principle for management and all serving on board ships. If that principle is to be translated into practice, a proper organisation is necessary, in order to ensure a consistent approach both to the care of the physical state of the ship and also to the manner in which it is operated. While the Master and the crew have direct responsibility for the technical and safety aspects of on-board ship operation, a department or suitably-experienced person ashore should be made responsible for those aspects from the shore standpoint.

1.2 Management — through the responsible department or person — should ensure that the following are all in order and should be familiar with the technical aspects of:

.1 the structure and stability of the ship, and the safety-related equipment on board;

.2 specialised equipment carried, particularly cargo-handling systems and navigational aids;

.3 documentation required to be on board, either because it attests that the ship is up to recognised standards (e.g. certificates of survey, crew certificates, etc.), or because it is necessary for the safe and proper operation of the ship (e.g. charts, guides, manuals). Care should be taken to ensure that documentation is up-to-date.

Where some of these responsibilities are delegated to the Master, management should give him full support in carrying them out.

1.3 Safety and operational policies should be clearly defined and publicised to all employees. They should be raised as a regular item for discussion both at management meetings ashore and at safety meetings on board.

2. SHORE-BASED PERSONNEL

2.1 Management should ensure that the relevant shore-based personnel:

.1 are aware of the basic technical aspects of the ship and its operation (as in 1.2) and are prepared to respond to the technical and operational needs of the shipboard personnel at all significant decision stages, e.g. from ship design/ordering to actual day-to-day operation;

.2 provide for a full and free exchange of information between shore and ship, particularly on any relevant navigational or operational matters, new technological developments, overall ship safety and personal safety;

.3 understand fully the implications of commercial decisions, in terms of the safety of the ship and the possible effect on the marine environment;

.4 make adequate provision for crew members' well-being e.g. proper accommodation and recreational spaces, proper catering arrangements, and medical care;

.5 regularly review procedures to ensure compliance with all the items in this Code.

3. SHIP-BOARD PERSONNEL

3.1 There should be a clear and planned approach to "personnel" matters concerning the crews employed on ships operated by the company. It is a direct management responsibility to provide ships with qualified and reliable seafarers and to give them additional training if required.

3.2 Specifically, management should ensure that the crew members:

.1 are sufficient in number to perform the tasks required of them, bearing in mind the basic principles and guidance contained in IMO Resolution A.481 (XII) and the need for proper duty/rest periods. (Allocation to specific tasks on board should remain the responsibility of the Master);

.2 are medically fit and have the requisite basic qualifications and experience in accordance with the Convention (STCW) and Resolutions adopted by the IMO Conference on the Training and Certification of Seafarers in 1978;

.3 have a proper knowledge of the technical aspects of the ship and its operation as necessary for the performance of their duties (as in 1.2);

.4 receive any necessary additional training, either in company procedures, or for familiarisation with the particular ship or equipment;

.5 continue at regular intervals to receive information, and where necessary training, in order to bring them up-to-date with new technological and other developments;

.6 maintain close communication with the shore-based personnel on any relevant navigational or operational matters;

.7 are provided with up-to-date navigational and other documentation in a language or languages fully understood by the crew;

.8 are regularly reminded of the need at all times for safe and clean ship operations, and for personal safety on board.

3.3 Where the Master finds that the points listed in 3.2 are not satisfactorily covered, for whatever reason, it is important that he take corrective action and/or raise the matter with management, as appropriate.

4. EMERGENCY PROCEDURES

4.1 It is important that the authority of the Master to take action in the event of an emergency involving the ship should not be compromised. Proper arrangements should be established which ensure an effective response to the incident, both by the crew on board and by the shore-based company organisation.

4.2 Management should ensure the development of:

.1 proper on-board emergency procedures, including regular and realistic drills;

.2 proper emergency back-up systems ashore, including an effective machinery for responding to the emergency;

.3 proper procedures to be followed both by ship and shore personnel concerning calls for outside assistance, including particularly the engagement of salvage services;

.4 reporting-back arrangements for all emergencies and near-emergencies;

.5 a system which will enable an incident to be assessed properly and any lessons to be learned.

4.3 Management and the Master should ensure that the procedures outlined in 4.2 are fully understood and adhered to.

5. COMMUNICATIONS

5.1 It is important that management, including senior management, regularly communicates with sea-going employees. Management representatives should visit each ship from time to time in order to review practices and procedures on the spot. Seminars and briefings for appropriate personnel might also be organised.

5.2 The objective should be to "motivate" sea-going employees by providing information in clear, digestible form on a regular basis — not just during a crisis. The information should cover company policy on safety and operating practice, and conditions of employment. It is essential for a climate of mutual trust to be built and maintained.

5.3 Management should develop effective two-way communication between shore-based and shipboard personnel; and should ensure that technical and company information passed to the ship is properly disseminated and reactions obtained.

6. GUIDANCE

6.1 In parallel with the growing number of regulations, an ever increasing amount of guidance to companies operating ships is becoming available in one form or another. This creates considerable difficulty for companies in keeping abreast of the paperwork which is published.

6.2 In terms of national legislation, management will need to be familiar with the relevant legislation and guidance in (1) the flag state and (2) states and ports visited by the ship.

6.3 Internationally, management should be familiar with the basic contents of the accepted "package" of international instruments. This includes such Conventions/Protocols as SOLAS, Load Line, MARPOL, Collision Regulations, ILO Convention 147, and STCW. A brief résumé of the various conventions and instruments — and their inter-relation — is given on the back page.

6.4 Also of direct importance to management is the guidance issued by national and international industry organisations, both in regard to general operational practice and to specific technical detail. These include technical guides concerning ship operations, navigational checks-lists, etc.

The Major International Shipping Conventions and Guidance

DEALING WITH THE SHIP . . .

SOLAS (Convention for the Safety of Life at Sea) 1974 and 1978 Protocol lay down a comprehensive range of minimum standards for the safe construction of ships and for the basic safety equipment (e.g. fire-prevention, navigational, life-saving and radio) to be carried on board. SOLAS also contains operational instructions, particularly on emergency procedures, and provides for regular surveys and certificates of compliance. Supplementary requirements, primarily concerning inert gas systems and steering gear, are laid down in the 1978 Protocol. As a complement to enforcement by the flag state, the Convention renders ships of a contracting party liable to specific control by authorities in the ports of other ratifying states. This may include detention of the ship.

MARPOL (Convention for the Prevention of Pollution on Ships) 1973 and 1978 Protocol contain measures designed to prevent pollution caused both accidentally and in the course of routine tanker operations by oil and oily mixtures, noxious or harmful cargoes, sewage and garbage. It sets out requirements for storing, treating and discharging these substances (including provisions related to segregated ballast tanks and crude oil washing systems) and for the reporting of spillages.

COLREG (Convention on International Regulations for Preventing Collisions at Sea) 1972 lays down the basic "rules of the road" governing traffic at sea, including rights of way, safe speed, action to avoid collision, procedures to observe in narrow channels and restricted visibility, and signals to be used to warn of manoeuvres.

Load Line Convention 1966 sets the minimum permissible free-board, according to the season of the year and the trading area of the ship; special ship construction standards are laid down in regard to watertightness.

DEALING WITH THE SEAFARER AND THE SHIP . . .

ILO Convention 147 (Merchant Shipping (Minimum Standards) Convention) 1976 requires Administrations to have effective legislation on safe manning standards, hours of work, seafarers' competency, and social security; and sets employment standards equivalent to those contained in a range of ILO instruments (covering e.g. minimum age, medical care and examination, accident prevention, crew accommodation, repatriation, social security, training). Parties also have to ratify SOLAS, the Load Line Convention, and COLREG. It allows an Administration to apply its provisions (including the power of detention) to any ship which calls at its ports, whether or not the flag state has ratified the Convention.

DEALING WITH THE SEAFARER . . .

STCW (Convention on Standards of Training, Certification and Watchkeeping for Seafarers) 1978 lays down extensive certification and qualification requirements (including syllabuses and sea time) for senior officers; all officers in charge of watches in the deck, engine and radio departments; and ratings forming part of a watch. All such seafarers will be required to have a certificate, endorsed in a uniform manner. It also specifies basic principles to be observed in keeping deck and engine watches and special qualification requirements for personnel on oil, chemical and liquefied gas tankers.

IMO Resolution A.481 (XII) (on Principles of Safe Manning) 1981 recommends all Administrations to issue their registered ships with a document specifying the minimum number and grades of qualified seafaring personnel required to be carried from the safety standpoint. It gives basic principles and detailed guidance to be observed by Administrations when assessing the safe manning of ships.

OTHER SAFETY CODES AND GUIDANCE . . .

In addition to the instruments described above, IMO has published other conventions, recommendations and codes, dealing with such matters as search and rescue, safety in container operations, and the characteristics and handling of different types of cargoes (e.g. bulk chemicals, dry bulk cargoes, liquefied gases, packaged goods, etc.). The ILO has issued codes of practice on safety and health at work, including accident prevention on board ship, at sea and in port; and also advice on medical treatment of seafarers (with the World Health Organisation).

Guides and check-lists are also published by various industry bodies, particularly the International Chamber of Shipping (ICS) and the Oil Companies International Marine Forum (OCIMF) — list available on request. They cover primarily specialised ship operations (e.g. tanker safety, safe handling of specialised cargoes, bridge procedures, etc.).

This Code is issued jointly by the International Chamber of Shipping and the International Shipping Federation. Any enquiries should be addressed to the Secretary, The International Shipping Federation Ltd., 30/32 St. Mary Axe, London, EC3A 8ET, Great Britain.

Appendix Two

(Chapter Thirteen Refers)

The following extracts from the position descriptions for a Master and Chief Engineer officer, from the Hain-Nourse Limited Sea Staff Position Descriptions Manual, are reproduced with the kind permission of the Peninsular and Oriental Steam Navigation Company.

It is noteworthy that these position descriptions were produced in 1970 at a time when Hain-Nourse had a mixed fleet of ships with traditional and general purpose crews. Deck and engine maintenance was still dealt with separately.

It was thus at a changing point between the old and new types of organisation referred to in Chapter Three and some of the titles, authorities and tasks, reflect this.

It is also noteworthy that the term "accountable" has a meaning more in line with the definition of "responsible" referred to in the introduction.

POSITION DESCRIPTION — MASTER

Purpose

To so manage the conduct of the voyage upon which his ship is engaged and the activities of those employed on board as to secure for the Company the maximum return upon its investment.

To provide leadership for the ship's company of a quality that will encourage each man to perform to the best of his ability and to develop to his fullest potential.

Authorities, working relationships and tasks

The principal activities associated with the position of Master of a ship owned, managed or manned by the Company are listed on the following pages. Under each activity are given the authorities, the working relationships and the tasks, applicable to that particular activity. It will be seen that the main activities only have been listed, those that in general make the largest demands on the Master's time, for it would clearly be unrealistic to attempt to prescribe the action to be taken by the Master in every possible situation that may arise on board ship. The Master will appreciate, therefore, that this Position Description is to be taken as a working definition both of his duties as generally accepted and of his position within the Company organisation and that the Company continues to look to him to exercise his initiative and discretion as circumstances may require for the furtherance of the current venture, the protection of the Company's interests and the safety of his ship and crew.

Definition: categories of authority

Authority for the Actions which a Master must take in the performance of the tasks required of him has, for the purposes of this Position Description, been divided into three categories.

CATEGORY I: The authority to act and to tell no one.
CATEGORY II: The authority to act provided that a named person or body of persons is informed that the action has been taken.
CATEGORY III: The authority to act after first requesting permission for the action from a named person or body of persons.

In short:

CATEGORY I: Act
CATEGORY II: Act and tell
CATEGORY III: Ask then act

Delegation

A Master may, at his discretion, delegate to a subordinate any or all of the work involved in the performance of an assigned task together with the necessary degree of authority, but the obligation to ensure the proper and timely completion of that task remains with the Master.

Accountability

The Chairman (i.e. the Chief Executive Officer) delegates that part of his Authority involved with the daily work of the fleet to the Fleet Manager. Masters are therefore accountable to the Chairman through the Fleet Manager, but have the right of direct access to the Chief Executive Officer if they consider the circumstances make it necessary.

ACTIVITY

1. Employment of ship

Working relationships

Accountable to:	The Chairman.
Has accountable to him:	Chief Engineer Officer, Chief Officer.
Works closely with:	Fleet Manager, Company's Agents, Charterers, Charterers' representatives.

Tasks

Subject, at his discretion, to considerations of safety of ship and life:

(a) To prosecute, with all possible despatch, the voyage upon which the Company has placed his ship.

Authority: CATEGORY II, Chairman.

(b) To fulfill the contract entered into by the Company in respect of his ship.

Authority: CATEGORY II, Chairman.

ACTIVITY

2. Carriage of cargo

Working relationships

Accountable to:	The Chairman.
Has accountable to him:	Chief Officer.
Works closely with:	Fleet Manager, Marine Manager, Insurance Manager, Company's Agents, Charterers, Charterers' representatives.

Tasks

(a) As "Employment of Ship", task (b).
(b) To take all action necessary to protect the Company's interests in connection with the carriage of cargo.

Authority: CATEGORY II, Chairman.

ACTIVITY

3. Navigation and handling of ship (including shifting ship, mooring, unmooring and drydocking)

Working relationships

Accountable to:	The Chairman.
Has accountable to him:	Chief Engineer Officer, Chief Officer, Second Officer, Third Officer, Radio Officer.
Works closely with:	Fleet Manager, Marine Manager, Engineer Manager, Insurance Manager, Pilots, Port Authorities.

Tasks

(a) To so navigate, handle and care for his ship (or arrange for the navigation, handling and care thereof) at sea and in port, as to:
1. Preserve ship and cargo from damage or loss by stranding, collision, contact, weather or other cause.
2. Ensure the safety of all persons on board.

Authority: CATEGORY I

(b) To assign navigating officers to bridge watches and to safisfy himself that any such officer is familiar with the equipment of the vessel and adequately briefed to effectively perform his tasks.

Authority: CATEGORY I

(c) To assign navigating officers to ship stations for mooring, unmooring, anchoring, shifting ships and similar operations.

Authority: CATEGORY I

(d) To ensure that upon undertaking any operation or manoeuvre such as shifting ship, mooring, unmooring, anchoring or drydocking, the ship is in all respects ready for such operation or manoeuvre.

ACTIVITY

4. Maintenance of ship's structure, equipment and machinery

Working relationships

Accountable to:	The Chairman.
Has accountable to him:	Chief Engineer Officer, Chief Officer, Purser/Catering Officer or Chief Steward, Radio Officer.
Works closely with:	Fleet Manager, Marine Manager, Engineer Manager.

Tasks

(a) To ensure that the entire ship's structure and all equipment and machinery on board are maintained at the standards prescribed by the Company, by statutory regulations and by the Classification Society.

Authority: CATEGORY I

(b) To arrange for the carrying out, when due, of statutory, classification society and other surveys and inspections.

Authority: CATEGORY II or CATEGORY III Chairman according to Company instructions.

(c) To engage the services of shore repairers or contractors to undertake work beyond the capabilities of the ship's resources.

Authority: CATEGORY III Chairman
In Emergency: CATEGORY II

ACTIVITY

5. Preparing for sea

Working relationships

Accountable to:	The Chairman.
Has accountable to him:	Chief Engineer Officer, Chief Officer, Purser/Catering Officer, Radio Officer.
Works closely with:	Fleet Manager, Marine Manager, Engineer Manager, Personnel Manager.

Task

To ensure that upon departure from any place the ship is ready in all respects for the intended passage, is seaworthy and is properly manned, equipped and supplied.

Authority: CATEGORY I

ACTIVITY

6. Storing

Working relationships

Accountable to:	The Chairman.
Has accountable to him:	Chief Engineer Officer, Chief Officer, Purser/Catering Officer or Chief Steward, Radio Officer.
Works closely with:	Fleet Manager, Marine Manager, Engineer Manager, Personnel Manager.

Task

To ensure that at all times the ship has on board for the intended voyage an adequate supply of stores and spare gear of all kinds and of fresh water.

Authority: CATEGORY I

ACTIVITY

7. Bunkering

Working relationships

Accountable to:	The Chairman.
Has accountable to him:	Chief Engineer Officer, Chief Officer.
Works closely with:	Fleet Manager, Company's Agents, Charterers Charterers' representatives.

Task

To ensure that at all times the ship has on board an adequate quantity of fuel for her intended service.

Authority: CATEGORY I

ACTIVITY

8. Ballasting

Working relationships

Accountable to:	The Chairman.
Has accountable to him:	Chief Engineer Officer, Chief Officer.

Task

To ensure that, as far as the ship's arrangement will allow, the ship has at all times a distribution of ballast and/or fuel and/or fresh water and/or cargo adequate to meet the requirements of seaworthiness and stability, having regard to the current state of weather and any manoeuvres or operations to be performed.

Authority: CATEGORY I

ACTIVITY

9. Administration, clerical and general

Working relationships

Accountable to:	The Chairman.
Has accountable to him:	Chief Engineer Officer, Chief Officer, Purser/Catering Officer or Chief Steward, Radio Officer.
Works closely with:	Fleet Manager, Marine Manager. Engineer Manager, Personnel Manager, Insurance Manager, Chief Accountant, Chartering Manager, Company's Agents, Charterers, Charterers' representatives. H.M. Consuls & "Proper Officers", Superintendents of Mercantile Marine Offices, British Shipping Federation Officials, Board of Trade Surveyors, Classification Society Surveyors, Dominion and Foreign Government Surveyors. Home Office Officials, Ministry of Agriculture and Fisheries Officials, Cargo Surveyors, Custom Officers, Immigration Authorities, Port Health Authorities. Port, Harbour, River and Canal Authorities. Salvage Association Representatives. P & I Club Representatives, Company's Solicitors, Shore Repairers.

Tasks

(a) Subject, at his discretion, to considerations of safety of ship and life, to comply with all instructions, standing orders and regulations of the Company.

Authority: CATEGORY I

(b) To communicate with the Company on all matters affecting the Company's intrest and in this connection to arrange for the upkeep of the documentation specified in Appendix I to this Position Description.

Authority: CATEGORY I

(c) To communicate as necessary with Company's Agents, Charterers and Charterers' Agents.

Authority: CATEGORY I

(d) To discharge the obligations in respect of ship and crew placed on the Master of a British ship by the Merchant Shipping Acts, the Oil in Navigable Waters Act, the Factory Acts, the National Insurance Acts, the Income Tax Act, Statutory Instruments, and, in the case of a ship with an Indian crew, by the Indian Merchant Shipping Act, and in this connection to arrange for the upkeep of the documentation specified in Appendix II to this Position Description.

Authority: CATEGORY I

(e) To discharge the obligations in respect of conditions of employment of crew placed on the Master by the provisions of the agreements of the National Maritime Board and in the case of a ship with a non British Crew, of similar agreements of the country from which they are engaged.

Authority: CATEGORY I

(f) To ensure compliance on the part of the ship and those employed on board with regulations (as they are relevant) made by Dominion and foreign governments, by port authorities and by other regulatory bodies, and to arrange for the preparation of relevant documentation.

Authority: CATEGORY I

(g) To ensure compliance on the part of ship and crew with regulations made by the company.

Authority: CATEGORY I

(h) To participate in the work of international agencies for the co-ordination of search and rescue at sea and weather reporting at sea, and to arrange for the upkeep of relevant documentation.

Authority: CATEGORY I

(j) In all matters affecting the working of ship and crew,

to provide to those accountable to him such adequate and timely information as they require for the effective performance of their duties.

Authority: CATEGORY I

(k) To pass to those accountable to him such information regarding the future employment of the ship and the movements of personnel as is necessary for the maintenance of good morale.

Authority: CATEGORY I

(l) To ensure that each member of the ship's company receives the training, advice and assistance necessary to enable to perform his duties effectively and safely and to develop professionally to his fullest potential; in particular, to ensure that company instructions on shipboard training are carried out.

Authority: CATEGORY I

(m) To satisfy himself that each officer on board understands his duties as specified by that officer's position description and by the Master's own instructions, to report the company any lack of ability or competence on the part of officers that he, the Master, is unable himself to correct.

Authority: CATEGORY I

(n) To maintain an interest in all matters affecting the shipboard, domestic and personal welfare of the ship's company and to recommend or take (within the limits of published company policy) any action that may be desirable in this connection.

Authority: CATEGORY II Chairman

(o) To report service of outstanding merit by any crew member.

Authority: CATEGORY I

(p) To promote, by means of personal example and instruction, correct conduct on the part of all members of the ship's company.

Authority: CATEGORY I

(q) To maintain equitable discipline among all members of the ship's company and in this connection to take, at his discretion, any disciplinary action sanctioned by law; to report to the company any misconduct that he is himself unable to correct.

Authority: CATEGORY I

(r) When appropriate, to offer the ship's hospitality in the company's or charterers' interests.

Authority: CATEGORY I

(s) To arrange for the carrying out of boat, fire and other emergency drills in accordance with statutory requirements and for the training of the ship's company in all safety procedures necessary for their protection.

Authority: CATEGORY I

(t) To conduct inspections of accommodation, food and water in accordance with statutory requirements.

Authority: CATEGORY I

(u) To train the ship's Chief Officer in the duties of a Master.

Authority: CATEGORY I

(v) To act as chairman of the ship's management committee.

Authority: CATEGORY I

(w) To take into his charge Admiralty documents and equipment.

Authority: CATEGORY II Chief Executive Officer.

POSITION DESCRIPTION — CHIEF ENGINEER OFFICER

Purpose

To manage the activities within his area of jurisdiction in such a manner as will make the maximum contribution to the profitable employment of his ship.

To provide leadership for all people under his supervision of a quality that will encourage them to perform to the best of their ability and to develop to their fullest potential.

Authorities, working relationships and tasks

The principal shipboard activities associated with the rank of Chief Engineer Officer of a ship owned, managed or manned by the Company are listed on the following pages. Under each activity are given the authorities, the working relationships and the tasks applicable to that particular activity. It will be seen that the main activities only have been listed, these that in general make the largest demands on the Chief Engineer Officer's time. It would obviously be unrealistic, in a working definition of a Chief Engineer Officer's duties such as this Position Description is intended to be, to attempt to prescribe the action to be taken in every possible situation that may arise on board ship. The Chief Engineer Officer will understand therefore that the company looks to him not only to perform the particular tasks specified in this Position Description but also in general to exercise his initiative and discretion as circumstances may require for the furtherance of the current venture and the protection of the ship's interests. It is of course his duty at all times to obey the lawful commands of the Master and to take whatever action may be necessary from time to time to secure the safety of his ship and those on board.

Definition: categories of authority

Authority for the actions which an officer must take in the performance of the tasks required of him has, for the purposes of this Position Description, been divided into three categories:

CATEGORY I: The authority to act and tell no one.
CATEGORY II: The authority to act provided that a named person is informed that the action has been taken.
CATEGORY III: The authority to act after first requesting permission for the action from a named person.

In short:

CATEGORY I: Act
CATEGORY II: Act and tell
CATEGORY III: Ask then act

Delegation

A Chief Engineer Officer may at his discretion delegate to a subordinate any or all of the work involved in the performance of an assigned task together with the necessary degree of authority but the obligation to ensure the proper and timely completion of that task remains with the Chief Engineer Officer.

ACTIVITY

1. Maintenance of machinery, equipment and machinery spaces, drydocking and repairs

Working relationships

Accountable to:	Master.
Has accountable to him:	All Engineer Officers, Electrical Officer.
Works closely with:	Company's Engineer Manager, Company's Engineer Superintendents, Company's Assistant Superintendent (Electrical) Chief Officer, Radio Officer, Shore Repairers, Contractors, Government and Classification Society Surveyors.

Tasks

(a) To develop and implement the programme of work required for the upkeep of the ship's machinery, equipment (including safety equipment) and machinery spaces in accordance with the company's current instructions on areas of officers' maintenance jurisdiction, standards of maintenance required and procedures to be followed.

Authority: CATEGORY II Master

(b) To ensure the safety of all persons engaged on maintenance tasks within his area of jurisdiction.

Authority: CATEGORY I

(c) To engage shore repairers or contractors to undertake work or provide services beyond the capabilities of the ship's resources.

Authority: CATEGORY III Master

(d) To ensure the proper execution of ship's work entrusted to shore repairers and contractors.

Authority: CATEGORY I

(e) To ensure the observance of safety precautions and relevant company and port regulations when shore repairers or contractors are employed on board.

Authority: CATEGORY I

(f) To progress classification society surveys on machinery and equipment within his area of jurisdiction in accordance with company's instructions.

Authority: CATEGORY II Master

(g) To keep up maintenance documentation required by the company and classification society.

Authority: CATEGORY I

(h) To advise the Chief Officer of any defect in ship's structure or equipment within the Chief Officer's area of jurisdiction that may come to his notice.

Authority: CATEGORY I

(j) Jointly with Chief Officer to ensure watertight integrity of ship's hull prior to flooding of drydock.

Authority: CATEGORY II Master

(k) To ensure that at all times the ship has on board for the intended voyage adequate spare gear for the machinery and equipment under his care and to indent for such spare gear as necessary.

Authority: CATEGORY II Master

(l) To ensure that all spare gear under his care is at all times properly stowed and kept.

Authority: CATEGORY I

ACTIVITY

2. Availability and functioning of machinery, equipment and services

Working relationships

Accountable to:	Master.
Has accountable to him:	All Engineer Officers, Electrical Officer.

Tasks

(a) To ensure that all machinery and equipment within his jurisdiction is available for use in the service of the ship as required by the Master.

Authority: CATEGORY I

(b) To arrange for the manoeuvring of main engines in accordance with bridge requirements.

Authority: CATEGORY I

(c) Subject to the authority of the Master to require action in the interests of the safety of the ship as a whole, to ensure that the machinery and equipment within his jurisdiction functions efficiently, safely and in accordance with builders', makers' and company's instructions.

Authority: CATEGORY I

(d) To assign engineer officers to engine room watch-keeping duties in accordance with his instructions and to satisfy himself that any such officer placed in charge of a watch is both technically competent and adequately briefed to control effectively the machinery and equipment in his charge.

Authority: CATEGORY I

(e) To inform the Master of any defect in, or condition of, ship's machinery, equipment or structure within his area of jurisdiction likely in any respect to affect the ship's operation or safety.

Authority: CATEGORY I

(f) To ensure that all domestic services provided by machinery and equipment within his area of jurisdiction are available as required for the well-being of those on board.

Authority: CATEGORY I

(g) To arrange for the upkeep of all logs, records and other documentation relating to the functioning of machinery and equipment within his area of jurisdiction and required by the company and regulatory bodies.

Authority: CATEGORY I

ACTIVITY

3. Ballasting and deballasting

Working relationships

Accountable to:	Master.
Has accountable to him:	Second Engineer Officer.
Works closely with:	Chief Officer.

Tasks

To arrange for the use of pumps, engine and pump room valves or other engine and pump room equipment for the purposes of working ballast in accordance with Chief Officer's written request.

Authority: CATEGORY I

ACTIVITY

4 Storing

Working relationships

Accountable to:	Master.
Has accountable to him:	Second Engineer Officer, Electrical Officer.
Works closely with:	Company's Engineer Manager, Company's Catering Superintendents, Company's Engineer Superintendents, Chief Officer, Suppliers.

Tasks

(a) To ensure that at all times the ship has on board adequate engine department stores and equipment for the intended voyage and to indent for such supplies as necessary.

Authority: CATEGORY II Master

(b) To ensure that stores accepted on board are of good quality and as intended for and to arrange for the issue of receipts to suppliers in accordance with quantities received on board.

Authority: CATEGORY I

ACTIVITY

5. Bunkering

Working relationships

Accountable to:	Master.
Has accountable to him:	All Engineer Officers.
Works closely with:	Chief Officer.

Tasks

(a) Jointly with the Master to decide the amount of fuel of various kinds to be taken on board from time to time (and to provide the Master with the necessary information to arrive at this joint decision) so that at all times the ship has on board an adequate quality of fuel for her intended service.

Authority: CATEGORY I

(b) Jointly with the Master to decide the distribution by compartments of the fuel on board that will from time to time best meet the requirements of seaworthiness, stability and economy of use and to arrange the transfer of oil on board or the loading of tanks necessary to achieve this distribution.

Authority: CATEGORY I

(c) To comply with company's current instructions on bunker receiving procedure and acceptable fuel specifications.

Authority: CATEGORY I

ACTIVITY

6. Administration, clerical and general

Working relationships

Accountable to:	Master.
Has accountable to him:	All Engineer Officers, Engineer Cadets, Engine and Deck Fitters, Engine Department Petty Officers and Ratings.
Works closely with:	Chief Officer, Purser/Catering Officer or Chief Steward.

Tasks

(a) To report to the Master any shipboard occurrence, condition or significant body of opinion of which the Master may be unaware and which may affect the efficient working or the safety of the ship or the well-being of those on board.

Authority: CATEGORY I

(b) To participate in the proceedings of the ship's management committee.

Authority: CATEGORY I

(c) To plan, in conjunction with the ship's management committee, the effective utilisation for all purposes of the manpower and other resources at the ship's disposal. (GP manned ships).

Authority: CATEGORY I

(d) To work together with and to assist the heads of other departments on board in whatever may be desirable for the good of the ship as a whole.

Authority: CATEGORY I

(e) To keep heads of other departments on board informed of events and developments within his own area of jurisdiction which may in any way affect the working of those departments.

Authority: CATEGORY I

(f) To pass to those accountable to him such information regarding the working and the movements of the ship as is necessary both for the proper performance of their duties and for the maintenance of good morale.

Authority: CATEGORY I

(g) To develop and implement a programme of shipboard training for engineer cadets in accordance with Company's current instructions.

Authority: CATEGORY I

(h) To train the ship's Second Engineer Officer in the duties of a Chief Engineer Officer.

Authority: CATEGORY I

(j) To assist engineer officers in their professional studies and to give them any advice or instruction they may require in order to perform their duties correctly.

Authority: CATEGORY I

(k) To maintain an interest in all matters affecting the shipboard, domestic and personal welfare of those accountable to him and to give advice, to enquire into complaints and to recommend any action that may be desirable in this connection.

Authority: CATEGORY II Master

(l) To promote, by means of personal example and instruction, correct conduct on the part of those accountable to him.

Authority: CATEGORY I

(m) To maintain equitable discipline among those accountable to him and to report to the Master any misconduct or lack of ability that he is himself unable to correct.

Authority: CATEGORY I

(n) To ensure compliance, in so far as those parts of the ship within his area of jurisdiction, and those accountable to him are concerned, with regulations made by the company, port authorities, classification socities, governments and other regulatory bodies.

Authority: CATEGORY II Master

(o) To keep up (or arrange for the upkeep and preparation of) log books, records, reports, performance data, correspondence and other documentation required by the company, charterers, port authorities, classification societies, governments and other regulatory bodies and which concern his area of jurisdiction and those accountable to him.

Authority: CATEGORY I

(p) To undertake any consultations with shore personnel that may be necessary in connection with the tasks specified in this Position Description.

Authority: CATEGORY I

(q) To instruct those accountable to him in safety procedures, to supervise the carrying out of fire and other emergency drills, and to participate in boat drills.

Authority: CATEGORY II Master

(r) To participate in inspections of accommodation, food and water required by statutory regulations.

Authority: CATEGORY I